ELECTROTECHNOLOGY

ELECTROTECHNOLOGY
Industrial and Environmental Applications

by

Nicholas P. Cheremisinoff, Ph.D.

np

NOYES PUBLICATIONS

Westwood, New Jersey, U.S.A.

Library of Congress Catalog Card Number: 96-28820
ISBN: 0-8155-1402-6
Printed in the United States

Published in the United States of America by
Noyes Publications
369 Fairview Avenue
Westwood, New Jersey 07675

10 9 8 7 6 5 4 3 2 1

Library of Congress Cataloging-in-Publication Data

Cheremisinoff, Nicholas P.
Electrotechnology : industrial and environmental applications / by Nicholas P. Cheremisinoff.
p. cm.
Includes index.
ISBN 0-8155-1402-6
1. Electricity--Industrial applications. 2. Environmental engineering--Technological innovations. I. Title.
TK4015.C44 1996
621.3--dc20 96-28820
CIP

PREFACE

A survey of electrotechnologies and their status is presented here. Principles of operation and significant applications both current and potential are outlined and an assessment is made wherever possible of the selected topics.

Many of the technologies and processes discussed are in their infancy and development stages. Some have developed and are developing rapidly; while all show great future promise. Rapid progress is being made in numerous industrial and environmental applications. The electrotechnologies identified in the volume have been selected for evaluation based on their potential impact in key industrial sectors and implications for industrial energy patterns.

There is little doubt that a "revolution" in electrotechnologies has been and is now underway in industry. In many cases far reaching as well as publicized developments have been fostering this revolution. Many of the implications are just starting to be realized in a wide range of industries.

The objectives of this book are two-fold:

- To identify and describe the range of industrial and environmental applications of electrotechnologies.
- To identify those applications that have potential for commercialization and that are likely to affect energy consumption patterns.

What follows in this book is a discussion of the key industrial sectors targeted, conclusions, and brief technical descriptions.

Nicholas P. Cheremisinoff

ABOUT THE AUTHOR

Nicholas P. Cheremisinoff is a private consultant to industry, academia, and government. He has nearly twenty years of industry and applied research experience in elastomers, synthetic fuels, petrochemicals manufacturing, and environmental control. A chemical engineer by trade, he has authored over 100 engineering textbooks and has contributed extensively to the industrial press. He is currently working for the United States Agency for International Development in Eastern Ukraine, where he is managing the Industrial Waste Management Project. Dr. Cheremisinoff received his B.S., M.S., and Ph.D. degrees from Clarkson College of Technology.

NOTICE

CONTENTS

1 ELECTROMAGNETIC INDUCTION HEATING

Electromagnetic induction heating utilizes a changing magnetic field to induce electrical eddy currents in the material being heated. These eddy currents dissipate energy, thus heating the material. This process is only appropriate for electrically conductive materials such as metals. The predominant means of creating a strong magnetic field is through the use of a coil. The flow of AC current through a coil generates an alternating magnetic field which extends through the material being heated. Coils are individually designed to heat simple or very complex shapes. The heating pattern, depth, rate and uniformity can be closely controlled through the coil design, power supply frequency and power density level. Thus, induction heating can be used for surface heating, through heating, or melting and is widely available for a wide variety of applications.

The largest use of coreless induction heating equipment is in molten metal melting, refining and holding.

Applications and typical products are:

Preheating:

- -- Forging of gears, shafts, handtools.
- -- Extrusion of shafts and structural members.
- -- Heading of bolts and other fasteners.
- -- Rolling of slab and sheets.

Heat Treating:

- -- Surface hardening and tempering of gears, shaft, machine tools, chain links, spring steel.
- -- Seam welding: of tubings and pipes.
- -- Melting: air melting of steel ingots, billets, castings.
- -- Vacuum melting: of steel ingots, billets, castings; nickel-based superalloys; titanium alloys.

Typically efficiencies are 60% as compared to fossil fuel furnace efficiencies of 25-40%. Reduced scaling and scrap losses. Extremely clean, quiet operation from an environmental standpoint.

Induction heating equipment has also been used increasingly for hot forming, forging, rolling, piercing and extruding.

Installed cost is typically three times that of a comparable gas furnace but because of higher efficiency, reduced scale and scrap losses, and half the labor requirements for operation and maintenance, often cheaper than gas furnace over the lifetime of the furnace.

Competing technologies include:

- Fossil fuel furnace, electric resistance furnace.
- Direct rolling, conventional welding.
- Ion nitriding, laser hardening, electron beam hardening, high frequency resistance hardening.
- Vacuum are melting, electron beam melting.

Other advantages:

- Fast start up and can be turned off when not in use; hence, saves energy.
- Heating is quick, which increases production.
- Since heat is generated internally within the workplace, size of the working area is substantially reduced because the need for a furnace enclosure, fuel delivery system, etc. is eliminated.

The greatest potential market is for large-scale machines for heat processing and metal treating. It is anticipated that the market share will increase appreciably in the future.

Generally, the large-sized and capacity machines require the use of considerably automated monitoring, handling and protection devices, while small-scale induction heaters may be entirely manually controlled. Thus, costs of inductor heating equipment vary greatly depending on size. Nevertheless, equipment costs may be estimated based on the operating frequency of equipment. Equipment costs are comparable or just slightly more expensive than alternative heating method equipment.

Generally, equipment for large-scale heating applications utilize the 60-cycle frequency for heating below curie temperature and 180 cycles for heating above. Energy usage in kWh for heating steel is given in Table 1. For example, if 2 tons/hr of steel are to be heated to 200°F using 60-cycle power, the number of kWh required over a 5-hr period would be:

$$2 \text{ tons } (2000 \text{ lb/ton}), 200°\text{F } (5 \text{ hr}) (5.24 \times 10^{-4} \text{kW}) = 2090 \text{ kWh energy}$$

The primary advantages of coreless induction heating equipment include:

- Dramatically reduced energy costs over fossil fuel alternatives.
- Reliance on a secure and predictable energy supply (electricity).
- Reduction in the scale and size of equipment providing savings of floor space.
- Large-scale iron heating, reduction of scale loss from 3% for fossil fuel furnace to 1% for an induction furnace.

For most applications, coreless induction heating is a capital-intensive technology, but due to its lower energy consumption it is economically advantageous over competing technologies, especially motor generators and fossil-fueled furnaces.

TABLE 1

COST AND EFFICIENCY DATA FOR CORELESS INDUCTION HEATING EQUIPMENT[a]

Frequency Range	Estimated Efficiency	$/kW	Energy Usage for Heating Steel 1°F (Specific Heat = 1.16)
60-cycle	60-70%	60-70	5.24×10^{-4} kW/(lb/hr)/day
180-cycle	50-60%	90-110	6.18×10^{-4} kW/(lb/hr)/day
960-cycle	45-50%	105-120	6.80×10^{-4}
3,000-cycle	45-50%	120-145	7.15×10^{-4}
10,000-cycle	45-50%	145-165	7.15×10^{-6}
450-cycle (RF)	40%	200-225	8.50×10^{-4}

[a]Costs include allowance for work-handling apparatus.

For some applications, newer, competing technologies are surfacing. For surface treatment of metals, for example, laser technology has been introduced and is cheaper than induction heating. Another alternative to induction heating, for nonconductive products, is microwave heating, and technology development on infrared heating is underway.

GENERAL PRINCIPLES

Coreless induction heating is being used for an increasing variety of industrial applications requiring heating or melting of conducting solid materials. Some current uses include:

- Forging: heating of metal before shaping.
- Melting: reducing metal or ores to a molten state.
- Soldering: joining of two separate parts by heating, usually by the introduction of a soldering metal.
- Annealing: heating to remove or prevent internal stress.
- Tempering: heating and subsequent quenching of materials to produce the desired state of hardness and elasticity.
- Bonding: joining of two separate parts through heating.
- Shrink-Fitting: joining of two separate parts by expanding an outside part, positioning an inside part and shrinking upon cooling to provide a tight joint.
- Coating: covering with a layer through heating and flow of material.
- Crystal Growing: careful temperature control allows growth of large, pure and stable crystals.
- Sputtering: evaporative disposition of metal on a surface.

These applications span a large spectrum of temperature and machine size requirements. Generally each application is sufficiently specific to require some specialized design and integration effort, especially in small-scale applications such as bonding or shrink-fitting; it is less true of large-scale forging or melting applications.

The principle of induction heating is illustrated in Figure 1. When a copper coil is wound around a conducting workpiece and an alternating current is passed through the coil, a magnetic field is established that also

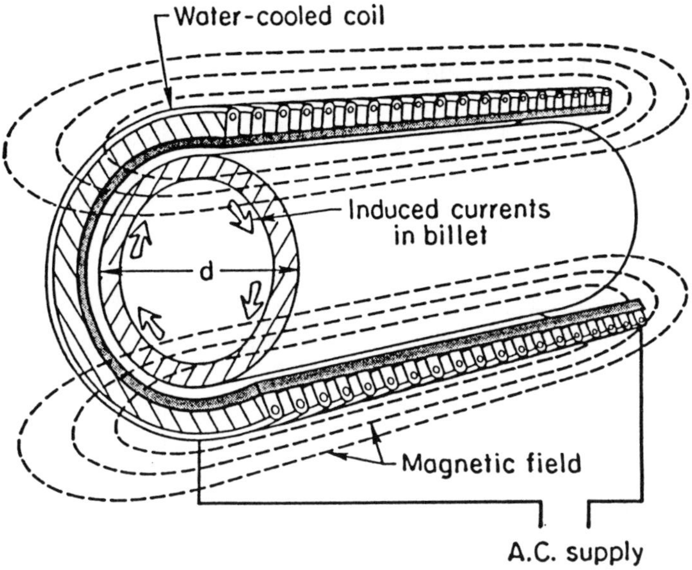

Figure 1. Induction heating for a tubular conductor heated by a solenoid coil.

causes a current to flow in the load. Linear induction coils enable the coils to correspond to the particular shape of the workpiece. The passage of current through the electrical resistance of the workpiece causes the workpiece to heat up. Distribution of the induced current through the load cross section is not uniform, and, consequently, heating is not uniform. The current decreases exponentially in magnitude from the surface to the center of the workpiece. The depth to which the current flows depends on the load resistivity, its permeability and the frequency of the alternating current. For steel, the two key factors are its magnetic properties and the AC power source frequency.

To obtain high heating efficiency, the diameter or cross section of the workpiece should be at least three times the heating penetration (the depth to which 87% of the induced heat is developed). Too high a rate increases the distance heat must flow and requires either a large heat time to permit heat to soak to the center, or a greater temperature differential between the surface and center of the workpiece.

EXISTING APPLICATIONS

Forging and Melting Applications

Induction heaters for forging and melting applications are generally large-sized coil or linear induction machines. For heating steel from ambient temperature to 2350°F in rolling mill preheating operations, induction heaters have been built up to a capacity of 600 tons/hr, requiring a power supply capable of delivering 200,000 kWh. Such large-capacity machines require considerable automation equipment for automatic handling of the production line and power monitoring to adjust the load power factor. In addition, sophisticated switching devices are required to switch the large power loads on and off without unbalancing or damaging the system. (Refer to Table 2)

In the rolling mill preheating application, the heater consists of four coils, each connected in parallel. An autotransformer can raise or lower the voltage of each individual coil ±10% in steps of 2%. This permits regulation of heat input to different sections of the lab to obtain uniform temperature. A circuit diagram for a 20-MW heater of this type is shown in Figure 2.

In Westinghouse induction heaters, regulation of heat input is achieved by varying the voltage by means of: (1) saturable reactors in series with the input powerline; (2) three silicon control rectifiers and associated firing controls, or (3) a dropper tube (vacuum diode) in the high-voltage d.c. line to the oscillator. A schematic diagram of these controls is given in Figure 3.

TABLE 2

FREQUENCY SELECTION CHART--HEATING STEEL FOR FORGING

Cross Section (in.)	(70-1300°F)	(1300-2200°F)
Over 6	60 Hz	60 Hz
4-6	60 Hz	180 Hz
2-4	60 Hz	1 kHz
1-2	180 Hz	3 kHz
½-1	1 kHz	10 kHz

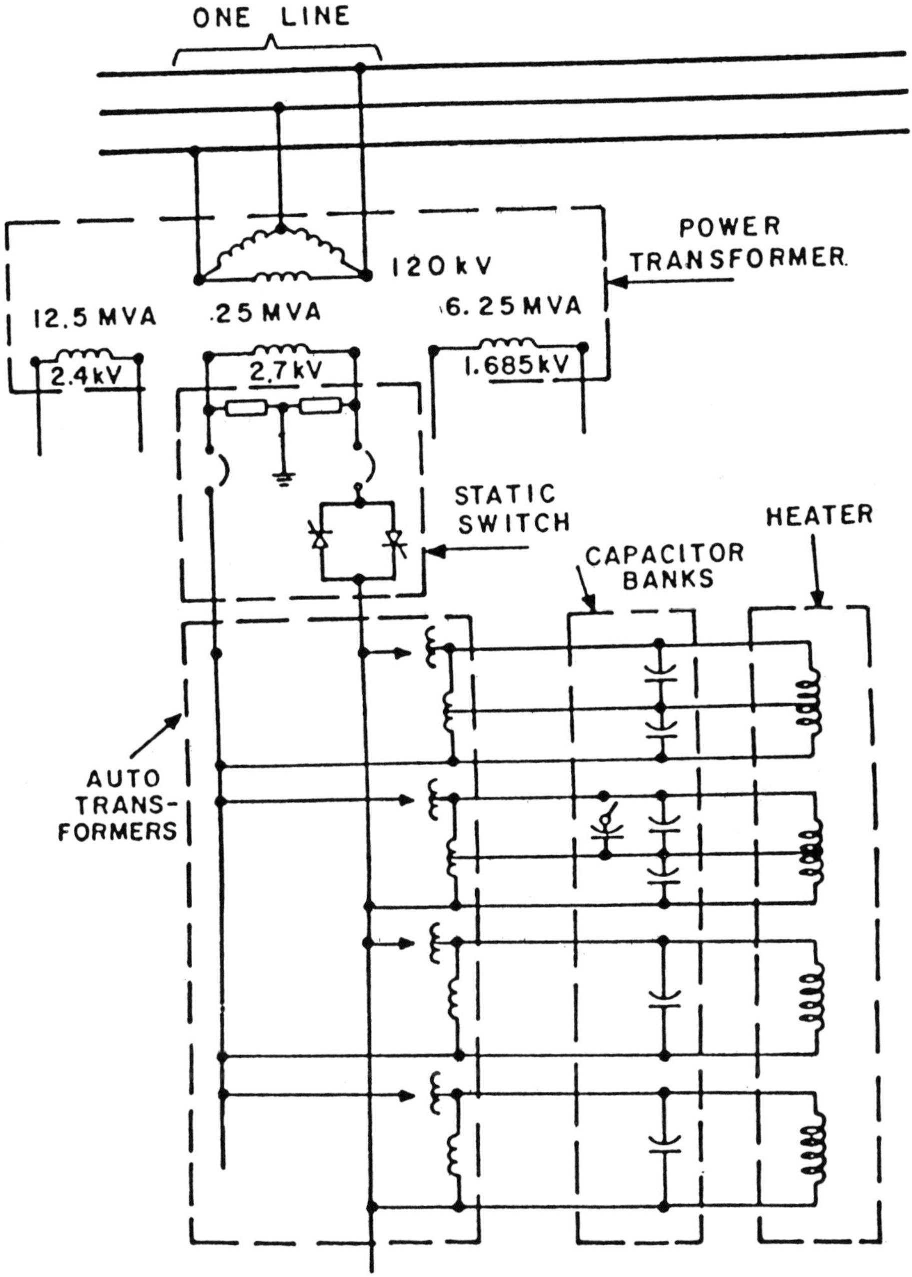

Figure 2. Electrical circuit for one of six heater lines.

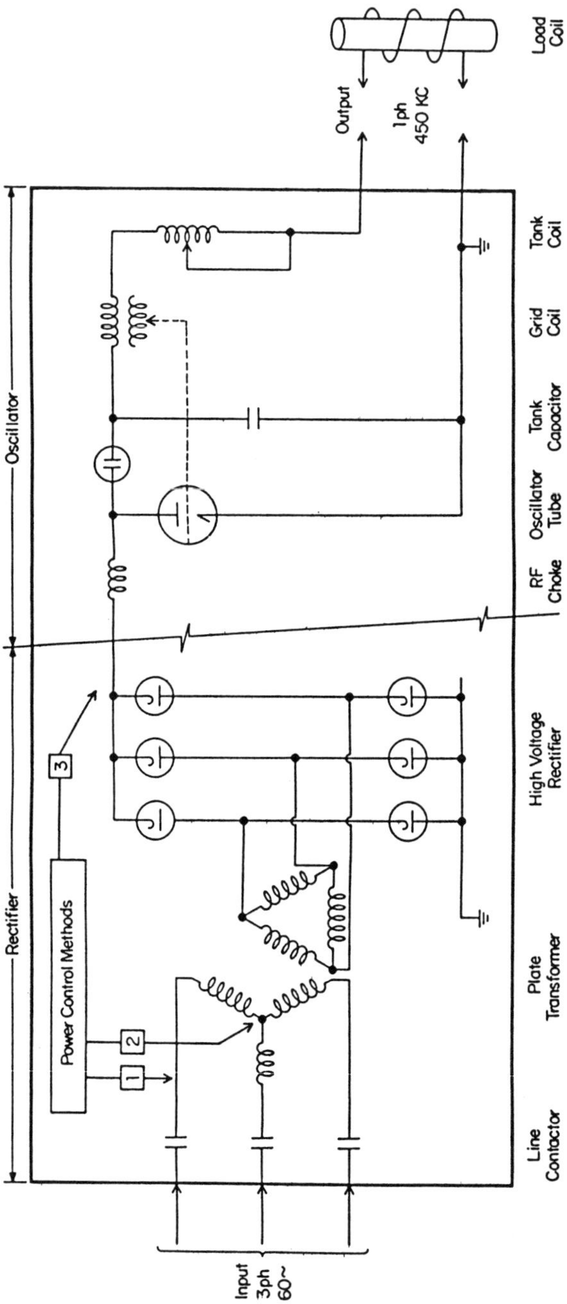

Figure 3. Typical schematic of Westinghouse power control mechanism.

The manual operation of such a rolling mill plant is inconceivable; no human being could operate such a vast and complex set of equipment without making serious mistakes. This specific installation includes five sets of control systems including slab handling control (digital); heater control (digital); static power switches (silicon control rectifiers); slab temperature control (analog); and process computer (digital). The controls during normal heating perform the following functions:

1. Slab Handling Control
   - Operates the gantry cranes.
   - Depiles slabs.
   - Charges heaters.
   - Deposits heated slabs on the mill, approaches table and sends them to the mill.

2. Heater Control
   - Changes tap settings and sets capacitor switches for each slab width.
   - Advances and retracts thermocouples.
   - Signals static switches to turn heaters on and off after checking the permissive circuits.

3. Static Switches
   - Switches currents on and off under lagging or leading conditions.
   - Detects and clears line-to-line faults.
   - Detects and clears line-to-ground faults.

4. Slab Temperature Control
   - By means of two proximity-type thermocouples per heater, slab surface temperatures are provided to the computer during the heating cycle. Low- and high-temperature adjustable limit switches are connected to recorder servos.

5. Process Computer

- Demand limit control.
- Phase balance control.
- Slab tracking and coil identification.
- Logging.

In addition to these systems, there are motor control centers, an annunicator system, a closed-circuit TV monitoring system, and an extensive protective relay system. Figure 4 shows the automatic handling and control scheme for the entire slab reheat plant.

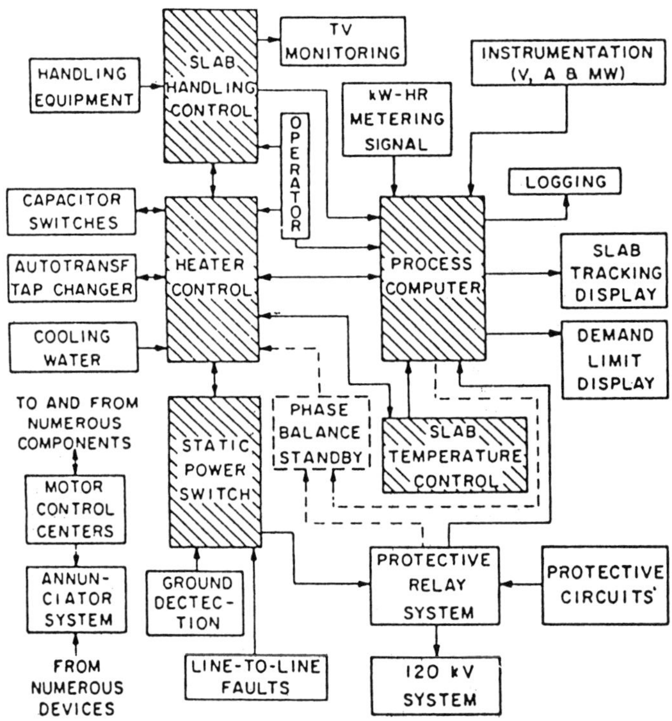

Figure 4. Block diagram of automatic handling and control scheme for slab reheat plant.

For melting applications, the heating rate is generally altered by regulating the power supplied to the coils as in the rolling mill applications. However, for one type of induction furnace, Pillar Corporation has provided for automatic monitoring of the power supply load. Virtually all other Pillar induction heaters use automatic monitoring equipment to regulate the heating rate through the power supplied to the coils. Pillar also utilizes more than one frequency to heat below and above the curie temperature (180 Hz, 640 Hz) for heating red brass and bronze to 2150°F. The switch from one frequency to the other is accomplished using solid state, serviconductor equipment.

Lindberg manufactures several lines of induction furnaces to melt and hold metal before casting. By using a two-chamber system, close temperature uniformity is maintained and charged metal has little effect on ladled metal. This results in fewer oxides and less sludge production. Ladling can be continued even while the channels are being rodded. This design also saves time on shutdowns to replace crucibles or to clean sidewalls, as required by other furnace designs.

Most large induction heaters are designed and constructed based on the specific application and part geometry required. Manufacturers of large-scale induction furnaces maintain design and applications staffs to develop induction heating equipment optimized for the application desired. These manufacturers include Ajax Magnethermic Corporation, Westinghouse Electric Corporation, Induction Process Equipment, Lindberg, and Pillar.

Technical data for a few standard induction furnaces for forging of steel are presented. The examples listed for Pillar and Induction Process Equipment Corporation represent the limits of capacity of the standard equipment detailed in the literature published by the above mentioned companies. Information on larger and smaller units is available upon specific request of these companies. Induction furnace specifications of the other manufacturers of large-scale equipment are also available upon request.

Power Requirements for Forging and Melting Applications

Capacity and power requirements of Pillar induction furnaces for melting applications are given. The furnace sizes listed are merely representative of the wide range of sizes and capacities available. Reference depth for common materials as a function of frequency is shown in Figure 6.

Manufacturers will supply technical data and costs of equipment for specific applications. Examples of currently operating large-scale forging equipment include:

- A pusher-type in-line slug heater installed for a major auto manufacturer which produces 8000 lb/hr of circular 4-in. radius by 2- to 4-in. long steel slugs. The five coils of the heater are supplied by 1500 kW of line frequency and 1500 kW of 180 Hz power.
- An installation to heat stainless and alloy steel billets for forging turbine blade preforms. The billets varied from 2.5 in. to 6 in. in diameter. A very precise and uniform temperature of 215°F was required. Frequencies of 60 Hz and 1 KHz were utilized with a combined power level of 750 kW.
- A long bar heater for hot shearing and forging railroad bearing races. It heats a 2-4 in. diameter by 20-ft. bar at a rate of 14,000 lb/hr. It is powered with 1250 kW of line frequency.
- A 10-ton furnace for melting with a capacity of 5200 pounds of aluminum. It uses 200 kW of input power to produce a melt rate of 1000 lb/hr at 1250°F. It operates at 240 KVA at a frequency of 60 Hz, nominal 230 V or 460 V.

Westinghouse has introduced a type of bar heater--the transverse walking beam heater. In this type of induction heater, bars are placed on walking beam rails. They are then literally walked through the heating line with the bar length transverse or 90°F to the direction of motor.

The walking beam concept provides a natural means of starting production with an empty bar heating line and provides continual production even with interrupted bar availability. When production stoppage is desired, bar feeding ceases to the infeed; in-process bars continue to heat and are fed directly to a forge press. If infeed bar interruption occurs due to availability or flaw during normal operation, an end-to-end bar forging machine must be shut down since a line stoppage without power removal results in melted bars. In the walking bar heater the bars may be walked out of the heater and fed directly to the press.

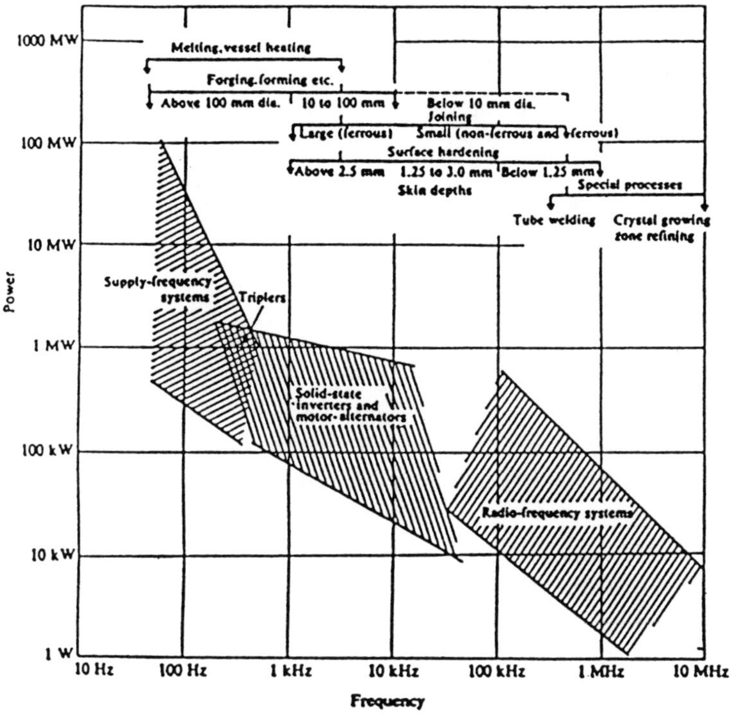

Figure 5. Power & frequency ranges for various types of induction heating generators.

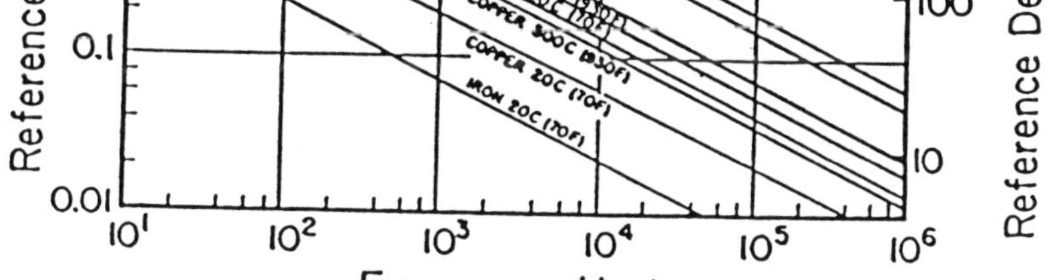

Figure 6. Reference depth for common materials as a function of frequency.

In the walking bar heater, the rails that carry the bars separate them so they are not in physical contact. This insulates the bars to prevent arcing and insures that parts do not weld together preventing handling problems and reducing scrap.

Cost Guidelines for Forging and Melting Applications

Since production requirements are known, the equipment costs can be estimated. These vary depending on the handling and auxiliary equipment required. However, typical costs for induction heating equipment for forging applications are given as a function of frequency and power requirements. For forging applications, the frequency of the power source is dependent on the diameter or cross section of the workpiece. Typical frequency requirements for steel workpieces are shown in Table 3.

To heat 6 tons/hr of high-volume 3.5 in. bars would require 2000 kW of power capacity comprised of 1000 kW at line frequency and 1000 kW at 1 kHz. The total cost for this package would be $250,000. However, the savings in energy cost that can be realized by introducing induction equipment more than compensates.

TABLE 3

EQUIPMENT EFFICIENCY AND COST

Frequency Range	Estimated Efficiency (%)	$/kW
60-cycle	60-70	60-70
180-cycle	50-60	90-110
960-cycle	45-50	105-120
3,000-cycle	45-50	120-145
10,000-cycle	45-50	145-165
450-cycle (RF)	40	200-225

NOTE:

1. Efficiency is overall thermal energy in work required divided by power from the incoming line.
2. Heating is from 70°F to 2150°F except in the case of 450 kilocycles equipment where maximum is 1200°F.
3. Costs include allowance for work handling apparatus.

Energy costs have typically comprised one-third to one-half the overall operating costs of heating for forging applications. When heating steel to forging temperature, average overall heating efficiency is approximately 65% in induction furnaces. This means 6 pounds of steel may be heated per kilowatt hour of electricity consumed or one ton of product would consume 333 kWh of electricity. Where high-frequency power supplies are used on smaller-sized billets and bars, these values are reduced by approximately 10-15% or to 5-5.5 lb/kWh. Estimates for the energy required to metal ferrous metals are shown in Table 4.

Although the energy input to an electric furnace is approximately half that of the cupola, if the energy consumed in generating the power is counted, the electric induction furnace above consumes, in effect, 5.49 million Btu, 1.5 times as much as the 3.64 million Btu consumed by the cupola.

A comparison of electricity costs for induction furnaces with fuel and gas costs for fired furnaces is given in Figure 7. The comparison is based on induction furnace efficiency of 65% and fossil fuel furnace efficiency of 20%. This efficiency indicates 3.5 million Btu are required per ton of heated product. This will vary depending on whether a more efficient closed-type rotary furnace is used or if the less-efficient

TABLE 4

ENERGY REQUIRED TO MELT FERROUS METALS

	Cupola	Electric Induction	Electric Arc
Heat Equipment Efficiency			
To preheat and melt to 2300°F	60%	60%	75%
To superheat from 2300°F to 2700°F	7	60	25
Millions of Btu to Preheat, Melt and Superheat 1 Ton of Cast Iron			
Theoretical	1.10	1.10	1.10
Actual			
Preheat, Melt and	1.60	1.59	1.28
Superheat	2.04	0.24	0.57
Total	3.64	1.83	1.85

open-type slot furnace is employed. The curve indicates electricity costs of \$0.1/kWh are equal to natural gas cost at \$0.90/million Btu or \$0.13/gal of heavy petroleum or refining residual fuel oil. By picking any point on the curve, one can readily determine how energy costs compare to this given set of conversion conditions. Where fossil fuel costs are above the curve for a chosen cost of electricity in cents/kWh, the fossil energy costs would be higher than using electric induction, conversely where they are below, fossil fuel costs would be lower.

Small-Scale Applications

Induction heating may be applied to a variety of small-scale applications including soldering, annealing, tempering, brazing, shrink-filling, bonding, crystal growing and sputtering. Induction heating is used to braze the heat to a generator housing to produce a high-strength, high-integrity joint for use in high-reliability aircraft and missiles. Rapid and localized induction heating permits soldering of parts containing heat-sensitive elements such as instrument cans, transistor seals and electronic components. It may also be used to bond knife blades to plastic handles. The load coil is positioned around the plastic handle and knife

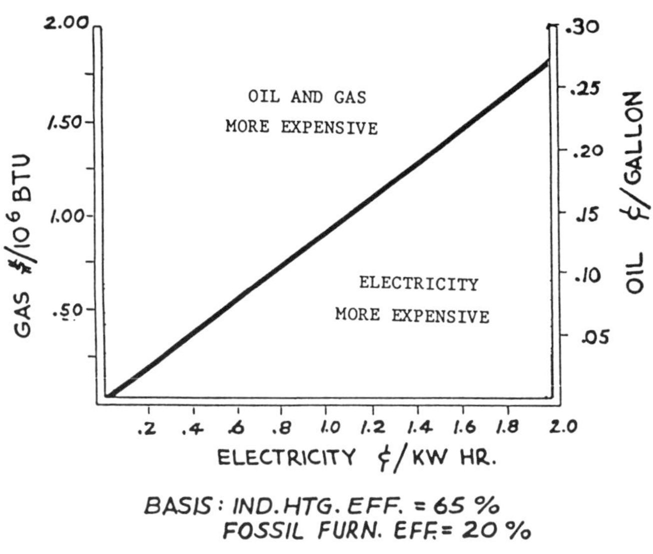

Figure 7. Heating for forging. Process heating energy cost comparison.

TABLE 5
GENERATOR SPECIFICATIONS FOR SMALL-SCALE APPLICATIONS

Generator Model	Power Input-KVA	3-Phase Voltage	R. F. Power Output (kW)	Frequency		Water Req. (gal/ton)	Cabinet Requirements (in.)		
				kHz	mHz		Wide	Deep	High
Lepel									
T10-T80	25.4-192	230,460,550	10-80	150-450	2.5-8	8-30	36-52	54-80	76
T1000-T200	240-480	460	100-200	150-450	2.5-5	35-60	52	80	76
T250-T500	590-1200	460	250-500	150-450	-	60-120	52	80	74
Cycle-Dyne									
AA series	1.9	120	1	450	27	5-50 psi	16	15	24
B series	45-120	200-250				3-6			
		400-500	20-60	450	3	gal/min	37	60	78
C series		200-250				3-6			
		400-500	5-15	450	3	gal/min	27	28	54
Leco									
521 or 523 series	1.0	115	-	-	14	-	16	30	24
621 or 623 series	1.0	220	-	-	14	-	16	30	24
763 series	1.0	115/220	-	-	14	-	16	30	24
Centorr									
600-6	-	-	30	4.2	-	variable flow--50 psi	-	-	-
Pillar									
1	4.2	230,460	3	max. 50	-	0.75 gal/min	24	18	56
2	8	230,460	5	50	-	0.75	19	30	42
3	16	230,460	10	50	-	1.5	19	30	42
4	40	230,460	25	50	-	3.5	34	24	75
5	80	230,460	50	50	-	7.5	34	48	75

blade. Only the metal tang is heated by induction, which in turn heats the plastic sufficiently to flow about the tang. A secure bond is formed without affecting the surface of the plastic handle or the previously tempered knife blade.

Shrink-fitting is another small-scale application of induction heating. For example, a cam shaft, collar and offset collar are heated to expand the inside diameter sufficiently so the shaft may be readily inserted. Glass-to-glass and glass-to-metal seals are also accomplished by induction heating by placing a preoxidized glass-coated Kovar ring between the pieces. The induction-heated ring softens the glass in the joint area and causes plastic flow to produce a seal.

Small-scale uses generally require specialized equipment designs for optimal process performance. Such equipment requires large initial investments in process equipment but the savings in energy, time and floor space can be considerable.

Manufacturers of induction furnaces for small-scale uses may be contacted individually regarding specific applications for design and cost estimates.

Typical examples of specific applications of the manufacturer's furnaces include:

- Equipment has been used for the selective hardening of gear sprocket teeth. Using a medium-carbon steel, the teeth of the sprocket reach 1600°F in 5 sec using the 10 kW inductor generator at a nominal frequency of 450 kHz. Equipment has been used in the assembly of multiple components in a single operation. A 7.5 kW generator has been used to produce two simultaneous sanitary brazes in 15 seconds in the assembly of a deep fat fryer.
- Leco equipment is designed solely to carry out quantitative analyses of carbon and sulfur content of materials. Complete carbon combustion is accomplished in 40 sec; complete sulfur combustion in 50 sec through the application of high temperature (1650°C) and high-pressure oxygen (to 20 psi).

Data regarding specific small-scale applications for equipment manufactured should be available upon request to the manufacturer.

POTENTIAL APPLICATIONS

Induction heaters may be used for heating or melting applications over a wide range of technologies. Solid state induction heaters generally may be used to replace motor generator heating equipment with a resultant increase in heating efficiency. Since induction heaters usually require large initial investments in equipment, some manufacturers have been reluctant to introduce induction heaters. The anticipated energy savings have made induction heaters increasingly attractive.

FUTURE ASSESSMENT

Major technological developments have centered on static power supplies and larger mass heating applications. Conventional rotating motor generators operating at frequencies up to 10 kHz have been largely replaced by solid state power supplies. It is expected that solid state device improvements, increased reliability and energy savings potential

TABLE 6

POTENTIAL USES FOR INDUCTION FURNACES AS A FUNCTION OF NOMINAL OUTPUT FREQUENCY

Frequency[a]	Application
80-200 kHz	Deep heating for hardening and forging plated parts.
180-400 kHz	Expitaxial growth; crystal growing; zone processing.
250-450 kHz	General purpose heating; surface hardening and joining operations.
2.5-5 MHz	Plasma processes; crystal growing; zone processing; heating thin parts.
15-30 MHz	Research and special applications.
30-50 MHz	

[a]Specific frequency will be determined by the generator, load coil and load being heated.

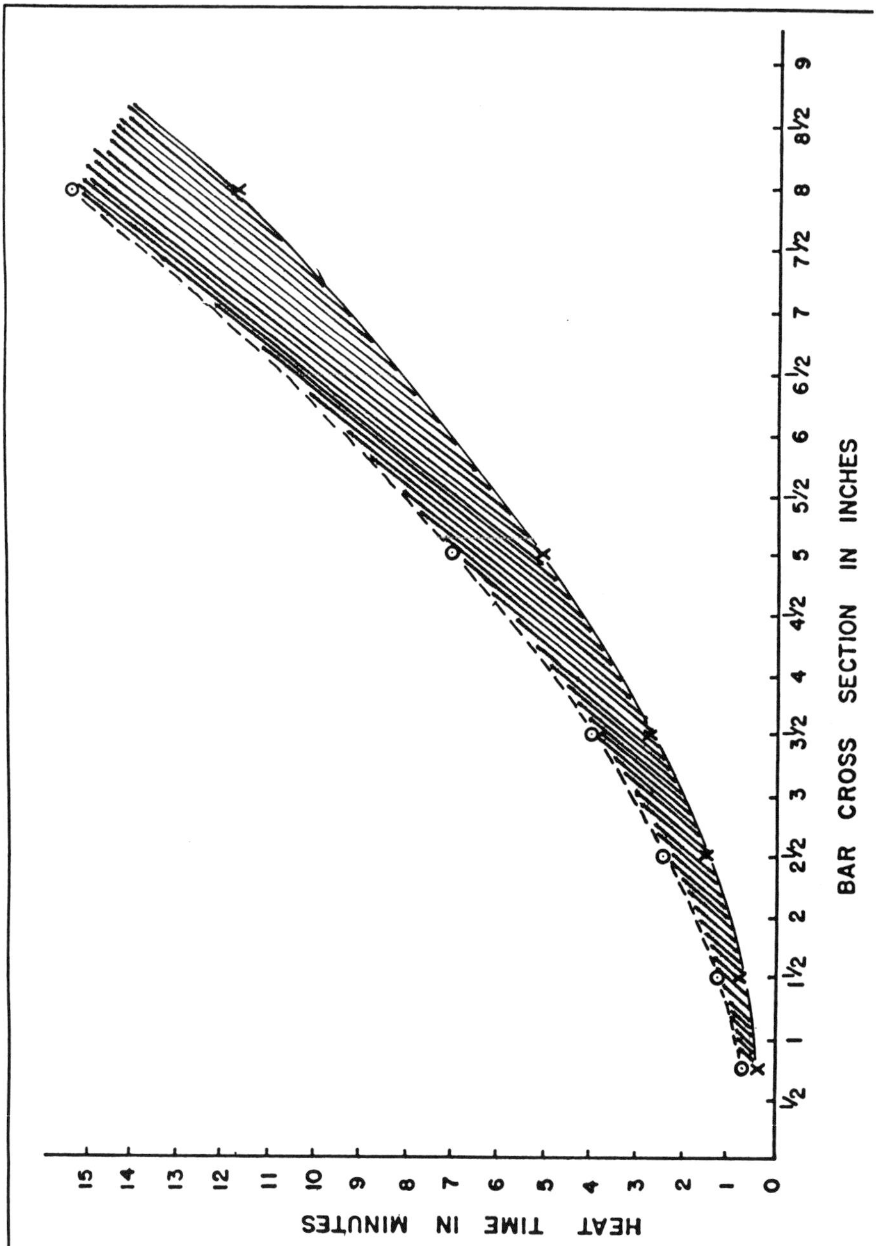

Figure 8. Heat time vs. bar cross section, through heating from room temperature to rolling temperature.

will advance sufficiently to convert many high-frequency heating to solid blooms, and on in-line processing applications and special furnace designs to reduce overall energy consumption. It will include developments instate. Emphasis will be on large mass heating such as slab, ingots and digital, programmable and computer-controlled applications to optimize heating. Induction melting will expand into larger ratings of 100 tons or more and several hundred-ton holding units to be used in conjunction with advanced-design automatic pressure paving machines.

Several technologies have emerged as alternatives to coreless induction heating. Infrared heating has centered primarily on low-temperature ovens for heating to 200-1200°F, for curing and drying applications. Higher temperature furnaces have been used selectively for heat-treating and even light-forging operations. Technology development in long-life, lower-cost, high-temperature elements is continuing.

Microwave process heating has been used primarily in curing or drying small-specialty nonconductive products, and operates at frequencies of 2000-3000 MHz, requiring Klystron and magnetron oscillators with waveguide transmission systems and requiring high capital equipment costs. Advances in solid state technology have allowed low- to medium-power units to be produced at reduced costs. Small, totally solid state, self-contained units for industrial applications will be developed.

For a variety of applications in heat treating metals, including guarding and hardening, laser beams are beginning to be used. It is currently cost-competitive and in some cases significantly cheaper than induction heating, carburizing and nitriding.

In heat treating, the laser beam irradiates the metal surface and is absorbed in an extremely thin surface layer. When the laser beam is moved rapidly across the metal surface, heat is conducted out of the original surface regions and into the base material. Thus, quenching occurs by conduction of heat into the interior of the metal.

2 RADIATION CURING
Ultraviolet, Infrared, Electron Beam

INTRODUCTION

Radiation curing of polymers involves the use of given wavelengths of light or electron beams to speed the curing mechanism, the dissociation of free radicals, from the minutes required in conventional curing to mere microseconds. Radiation curing is used for coatings on wood stocks, motor vehicles, metal and plastic containers, and for inks used for packaging, printing and specialty items.

Advantages of radiation curing for the two largest markets include:

- Rapid drying speeds (seconds or less).
- Reduction or elimination of organic solvents, thus eliminating air pollution and incineration problems.
- Significant reduction or elimination of fossil energy-heated drying ovens and incinerators.
- Coating of heat-sensitive materials (plastics).
- Increase in production rates.
- More efficient use of polymeric coating materials because of less penetration of flowing material into substrates.
- Savings in space of application equipment.

There are several disadvantages to radiation curing. Coating equipment upstream of the curing equipment, for example, may not be compatible or may negate the time advantage of radiation curing. Radiation does not travel around bends, thus requiring complex configurations of curing equipment for unusual three-dimensional shapes. Also, the chemistry has not been perfected, thus some formulations are severely limited in application. Some present coatings have mediocre chemical and weathering resistance, flexibility, abrasion resistance and poor film adhesion to impervious substrates. The technology itself poses certain problems, including the use of toxic ingredients, fears for radiation effects on workers and radiation damage to certain substrates. In addition, there is certain universal psychological inertia to accepting any new industrial process.

Radiation curing includes three technologies: ultraviolet (UV), infrared (IR), and electron beam (EB) curing. In UV curing the material being processed, the substrate, is coated with a radiation-curable formulation, then passed through a drying oven containing UV lamps. The medium-pressure mercury lamps dominate this market, although low- and high-pressure mercury lamps also can be used. The most popular unit is the 200 W/in., medium pressure, mercury-argon bulb with a claimed 2000 hr operating life and 63% radiation output-power input efficiency. Costs are about $250 per lamp. Even with multiple UV lamps being required in each dryer, plus safety features on the UV curing oven to keep extraneous radiation from escaping, total unit cost for UV curing is only several thousand dollars per production line.

Introduction of *infrared curing* has further lowered lamp costs. IR bulbs cost less than $100 each, and IR radiation that escapes is much less of a hazard to workers than UV. In addition, standard solvent-based and/or water-based coatings can be cured by IR.

Electron beam accelerators (EB), on the other hand, are very high in capital cost--on the order of $200,000 to more than $1,000,000 per production line. Nevertheless, EB accelerators, providing several thousand times the radiation energy provided by UV lamps, can do some things ultraviolet and infrared cannot.

Existing applications of radiation curing have resulted in a sizeable market for the technology. Totals came from three major market applications: wood coatings; containers, closures and metal decorating; and automotive or motor vehicle.

U.S. consumption of UV-cured inks is divided between packaging inks and all other printing. There is no competitive market for radiation-cured inks; its biggest drawback being its high price.

The energy savings obtained by radiation curing are based on the difference between the costs associated with the length of time the classical materials have to be heated at a given temperature less the costs of installing and utilizing the radiation-curing techniques. Further economic savings are obtained by the much smaller-sized equipment required for radiation curing, plus the greatly increased speed at which coated materials can be processed, thus lowering the capital and operating cost per unit of production. A comparison of energy consumption and costs between radiation curing and conventional techniques is given in Table 1. Most radiation curing is done by ultraviolet.

TABLE 1

COMPARISON OF ENERGY CONSUMPTION AND COSTS, RADIATION VS. CONVENTIONAL CURING

Application	Radiation Curing	Conventional Technique	Savings	Remarks
Coatings				
Metal cans	6.6 Btu/can (UV curing)	23.0 Btu/can (thermal process)	16.4 Btu/can	
Curing of polyethylene cable insulation	2.5¢/lb (by electron beam)	6.3¢/lb (using peroxide)	3.8¢/lb	Includes total costs
Curing of PVC extruded onto copper wire	$0.50/lb	$2.00/lb	$1.50/lb	Includes total cost of producing wire
Inks				
Printing Inks	$2.91/lb (UV curing)	$1.71/lb (thermal)	246% less energy than conventional methods	
Pressure-Sensitive Adhesive Tape	$13,000	$130,000 (thermal oven)	$117,000 in capital costs for a 300 ft /ft/min unit	EB curing capital cost would be $200,000

GENERAL PRINCIPLES

Technology

Polymers that cure by free-radical mechanisms are the basic raw materials for radiation curing. Classical curing of these polymers has been accomplished by heat or by absorption of sunlight. Both processes dissociate the contained free-radical source into active free radicals, which initiate polymerization of the curable polymer. The speed at which free radicals are formed and their concentration, determine the properties of the cured polymeric material.

Both classical approaches to free-radical source dissociation require rather lengthy periods of time; in addition, the thermal heating is energy inefficient. Use of ultraviolet, electron beam or infrared energy, however, can greatly speed the formation of free radicals from the source.

Radiation curing utilizes the principle of selecting a given wavelength or sequence of wavelengths of light or electron beams, which will speed the dissociation of the free-radical source from minutes to microseconds. The material being processed, the substrate, is coated with the radiation-curable formulation by knife edge, rollers, screens, etc., then passed through a drying oven that contains UV or IR lamps, or accelerators.

Polymer systems most widely studied and developed to date for radiation curing include: acrylates, polyesters (cross-linked with styrene), thiols (cross-linked with unsaturated compounds), epo epoxies, urethanes, vinyls and various mixtures of these polymers with monomers. Typical free-radical sources (photoinitiators) are incorporated into these polymers, along with pigments and photosensitizers, which act to increase the rate of dissociation of the free-radical source when the formulation is exposed to the proper radiation.

Because the technology currently provides a flowable liquid formulation that can be converted quickly into non-flowing, adherent solid, major applications of radiation-cured polymers developed to date are as coatings and inks. Because of the speed radiation curing, production times for applying these coatings and inks can be greatly reduced, thus greatly reducing unit cost of production per unit of capital investment. Curing times and characteristics of radiation-cured polymer formulations can be changed by changing the balance of photosensitizer, the pigment, incorporation of extenders, viscosity improvers, or even the use of small amounts of solvents.

Early radiation-cured formulation photoinitiators were also sensitive to atmospheric oxygen. Consequently, surfaces could be depleted of some of their constituents by atmospheric oxygen, giving rise to incomplete surface cures. To avoid this situation, equipment was modified to provide a blanket of inert gas (usually nitrogen). This technique adds additional capital and operating cost to the process. Modifications in polymer formulations provide formulations that are no longer sensitive to atmospheric oxygen, and obviate the need for inert gas. Use of inert gas with UV curing, however, has the advantage of eliminating the undesirable generation of ozone, and increases the amount of radiation energy available for production. This, in turn, can increase the speed of production.

Curing Equipment

Most commercial radiation-curing equipment was based on *ultraviolet radiation*, with the balance being *electron beam* accelerators.

Infrared (IR) radiation processing utilizes processing infrared wavelengths to heat surfaces. A high temperature heat source (or emitter) is generated so that the energy radiated by the source is of specific wavelengths which are most readily absorbed by the surface of the product, thus heating and drying or curing it. The speed and depth of drying are functions of the heat source temperature and the absorptivity of the material being dried, particularly at the specific IR wavelength. IR radiation is used for both drying and curing surface coatings, as well as for some heating applications. The main advantage of IR processing over convection heating are improved energy efficiency, higher production rates, space savings, precise control, lower maintenance, and overall improved product quality.

Infrared (IR) waves are the part of the electromagnetic specture between visible light and radio waves. IR wavelengths range from 0.8 to 1000 micron (μm) but the useful range for industrial heating is about 1 to 10 μm. Those radiant sources which emit energy at wavelengths nearest the visible light bank ($\sim$ 0.8 to 2 μm) are called "near infrared;" while those at the other end of the useful range (3 to 10 μm) are called "far infrared." Infrared energy is emitted from any body which is warm

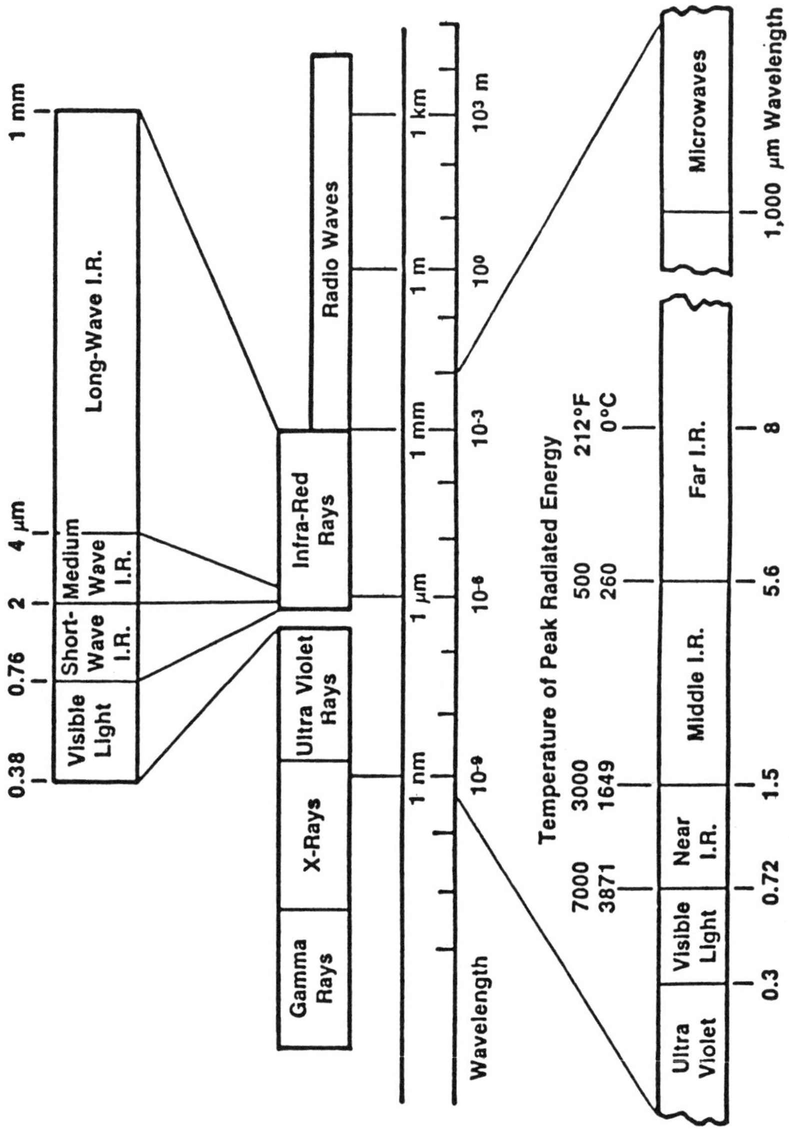

Figure 1. Electromagnetic spectrum.

(i.e., warmer than absolute zero). It is produced primarily by thermally induced vibration of molecules in solids, liquids, and gases, in contrast with ultraviolet radiation, which is primarily associated with electron excitation in gases.

All IR emitters produce radiation over a range of wavelengths, but the wavelength at which radiation is maximum is determined by the temperature of the emitter. An ideal emitter, called a blackbody, would emit radiation which is a plot of radiation energy against wavelength for different source temperatures. Each solid curve represents the radiant flux or density that would be emitted for various wavelengths if the source were at the specified temperature. The dotted line connects the peak radiation values for each temperature. From the graph it can be seen that as the temperature of the sources increases, the corresponding peak wavelength decreases and the peak radiant flux increases. Thus, a high temperature emitter provides short-wave radiation with high intensity, but as the temperature is reduced, the radiation becomes increasingly long wave and of lower intensity. It can also be seen that higher temperature emitters provide proportionately more of their energy near their peak wavelengths while lower temperature emitters have a broader energy distribution.

Actual emitters are less than ideal blackbodies (they are called graybodies) and have energy levels which are less than a blackbody. But regardless of the source of heat and the material from which the source is constructed, any emitter at a given temperature will have the corresponding wavelength characteristics. Therefore, if different types of infrared sources operate at the same temperature, they will tend to have the same peak wavelength as well as other characteristics such as penetration and color sensitivity.

Materials used for IR radiation are measured by their emissivity. Emissivity is the ratio of the total radiation emitted by a graybody surface to that of a blackbody at the same temperature. Emissivity values range from 0 to 1. An emissivity closer to 1 means that the surface behaves like that of a blackbody (i.e., it is a good emitter).

Radiation energy impinging on an object is partially absorbed, partially reflected, and sometimes partially transmitted. The degree of absorbance, reflectance, and transmittance depends on both the wavelength of the radiation and the bulk, physical, and surface properties of the target material. The radiation that is absorbed by the material is converted into heat, whereas the radiation that is reflected by or

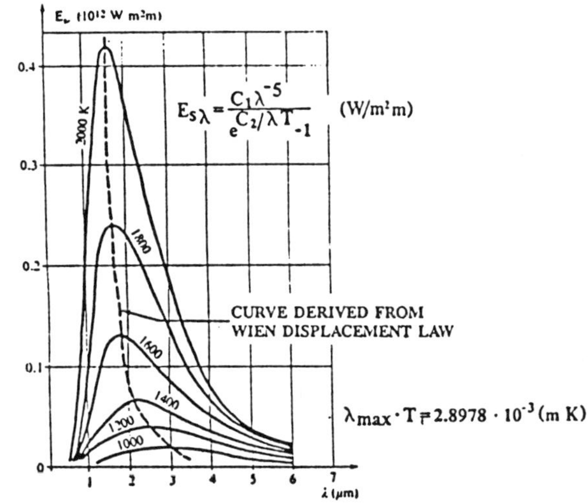

Figure 2. Wien Displacement Law: Relationship of IR emitter temperature to maximum intensity wavelength.

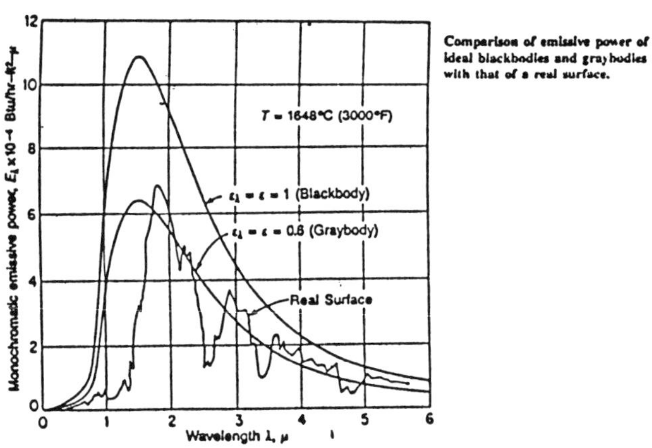

Figure 3. Radiation spectrum for different types of bodies.

transmitted through the material is "lost" energy unless it is reflected back onto the material or reabsorbed by the emitter to help maintain its temperature. Although a difficult task, it is possible to determine the absorptivity, reflectivity and transmissivity of a material at various wavelengths. Materials are typically selective, meaning that they absorb some frequencies better than others. In addition, the material thickness also influences absorption and the selectivity of the material will vary with depth.

Infrared radiation emitter processing equipment can be divided into two general categories: gas-fired units and electric resistance units.

Gas-Fired Infrared Sources

Gas-fired IR heaters are medium temperature sources which typically use direct fire refractory burners. There are also ceramic-faced burners with tiny nozzles known as porous refractory burners. Air and gas are mixed in the burner head or in a pre-mix chamber and burned at the ceramic face, which heats and radiates. Temperature of the ceramic is typically 1400 to 1600°F with a corresponding wavelength of 2.5 to 3.3 microns. Radiant efficiency is generally 30 percent but some burners can reach 60 percent.

Electric Infrared Sources

Electric infrared emitters have wavelengths corresponding to each of the three infrared radiation spectrum bands (short [near], medium, and long [far]). The temperature of short infrared emitters is greater than 2500°F, that of medium infrared emitters is between 850 - 2500°F, and that of long infrared emitters is less than 850°F. The various types of emitters all use the thermal effect of an electric current flowing through a resistive element (Joule effect). For this reason electric IR heating can be classified as a special form of radiation heating and resistance heating (since it is based on this energy transmission method and uses electric resistances as radiation emitting sources).

Short Infrared Emitters

Short infrared emitters consist of an evacuated tube or lamp, or more often, a lamp containing an inert atmosphere (argon, nitrogen) in which a tungsten filament is heated to a very high temperature (3650 to 4550°F). The maximum monochromatic emittance is around 1.2 μm. Approximately 5 percent of the radiation is in the visible wavelengths, which explains the bright yellow color of these emitters.

Infrared lamps, which are very similar in design to light bulbs, have a glass envelope which sometimes incorporates an internal or external reflector (the internal reflector is formed by employing an inside deposit of gold, silver or aluminum). Power required for each bulb is low, generally 150, 250, or 350 watts. The tungsten filament temperature is raised to 3650°F, which corresponds to a maximum emission wavelength on the order of 1.4 μm.

Infrared tubes consist of a quartz tube filled with an inert gas. The temperature of a spiral wound tungsten filament, supported by disks, is raised to about 4000°F. Quartz is practically transparent to infrared radiation, absorbing only about 5 percent of the energy. More than 50 percent of the absorbed energy is re-emitted in the form of IR radiation at a longer wavelength. Quartz is only slightly sensitive to thermal shock, because it has a very low coefficient of thermal expansion, offers adequate mechanical strength, and is a poor conductor of heat. It is for this reason that this material is widely used in the manufacture of infrared emitters. These tubes are available in different effective lengths (8 to 60 in.). The power output of an individual tube can vary from 500 to 7000 watts or even higher; tubes of 20 kW are available for special applications. For higher power density emitters, the bases and mountings are usually air cooled or, in some cases, even water cooled.

Very high power density tubes exist in which the tungsten filament temperature reaches 4900°F. To prevent evaporation of the filament, which could cause blackening of the tubes and diminish its efficiency and service life, a halogen gas (generally iodine) is added to the inert gas filling the tube. At this temperature the radiant output (efficiency) is around 86 percent. The phrase "high intensity infrared" is used by some manufacturers to describe heating arrays that produce heating energy of at least 100 W/in^2. This power density can only be attained with the short-wave IR devices.

Medium Infrared Emitters

Emitters for medium wavelengths generally operate in the range of 1300 to 2400°F. The usual emitter materials are nickel-chromium (nichrome) or iron-chromium-aluminum. These emitters are mounted in glass or quartz tubes, silica or quartz panels, or surrounded by metallic radiant tubes. Approximately 1 percent of the energy emitted by these devices is in the visible light range, giving them a light red color.

Single or double, clear or translucent silica tubes behave as a support for a resistance coil element which in most cases is an iron-chromium-aluminum alloy heated to a temperature of 1850 to 2450°F. The tubes can be goldplated at the back or use separate reflectors. A wide range of useful wavelengths (0.2 to 3 μm) and powers (250 to 8000 watts) exists. Cooling is not normally required for emitters of this type. The difference between these tube lamps and the short IR tubes is that the incandescent wire does not need to be protected from the air because of the alloy used and the temperature involved. Another medium-temperature source is silica or quartz panels using nickel-chromium (nichrome) or iron-chromium-aluminum filaments at temperatures of 1300 to 1850°F. Power varies from 800 to 1600 watts for an effective area of 100 in^2 (8 to 16 W/in^2). High specific quartz panels up to 32 W/in^2 exist.

Another medium-temperature source is a nichrome wire coiled and embedded in magnesium oxide and surrounded by a metal tube (generally a refractory stainless steel). The magnesium oxide is both a good electrical insulator and a good conductor of heat. The electric resistance is used to heat the tube by conduction. The radiation source is therefore not the filament but the metal sheet which emits at a temperature from 1300°F to a maximum of 1470°F. A large part of the energy radiated is in the long infrared; hence, these heaters are sometimes classified in this category. A common operating temperature for this type of IR heater is 950°F. To increase efficiency, these elements are generally installed in a reflector (reflectors greatly help all IR elements except panel heaters). Lamp-shaped emitters of this type exist in which the tube is spirally wound in one plane with a conical reflector used to concentrate the

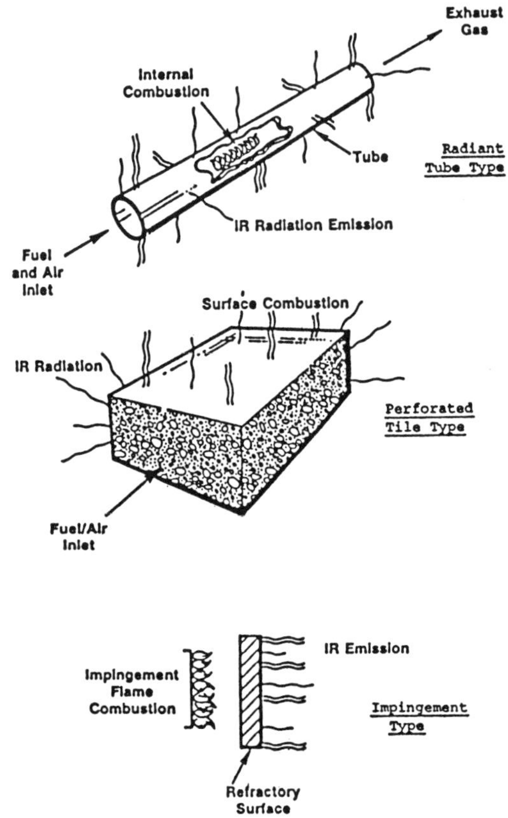

Figure 4. Common IR radiation emitters - fuel burning types.

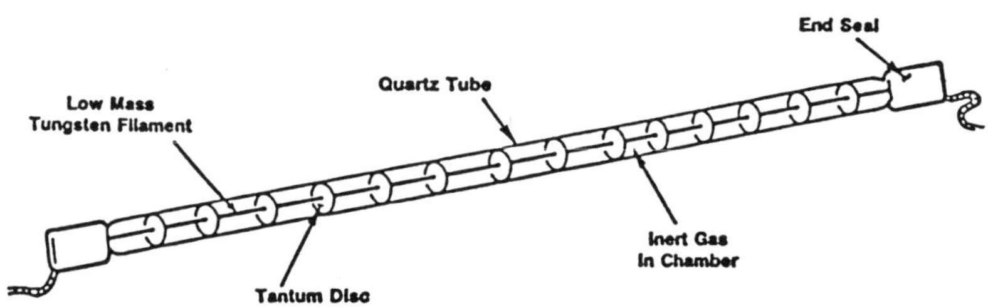

Figure 5. Electrically activated quartz tube IR emitter.

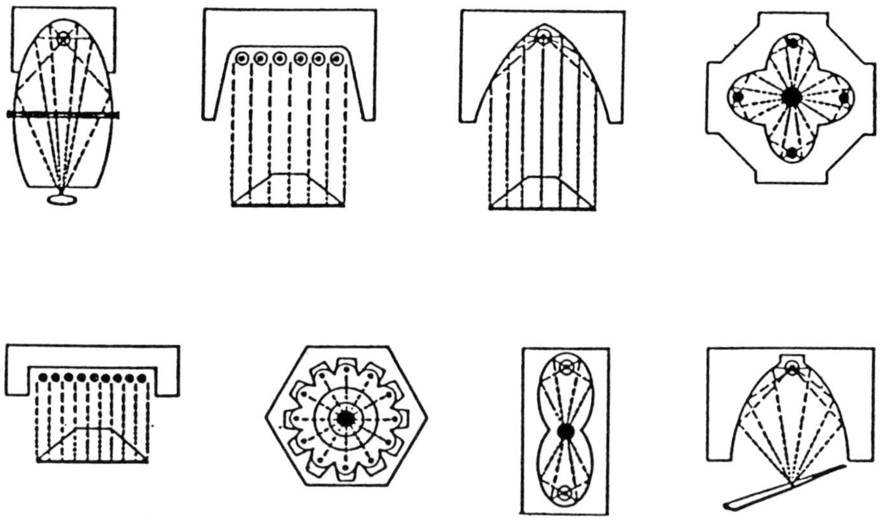

Figure 6. Reflector patterns.

radiation. These elements can also be shaped as required to fit the contour of the part to be heated. There are ceramic IR heaters powered by electric resistance nichrome wire. These fused ceramic elements operate at about 1000°F and emit a wavelength of 4 to 5 microns with a radiant efficiency of 40 to 45 percent.

Long Infrared Emitters

Long infrared emitters consists of either glass radiant panels which have been rendered electroconductive on the surface or vitrified ceramic covered panels. They are heated between 550 and 1100°F and sometimes as high as 1300°F. These sources do not radiate in the visible range.

Electroconductive radiating panels consist of a plate of hardened glass. The inside surface of the glass is coated with a thin layer of a metal oxide which is utilized as an electrical resistor to heat the glass. An aluminized sheet metal reflector and a glass wool insulator are located on the back surface. The permissible surface temperature depends on the type of glass used: 175°F for ordinary glazing glass, 300°F for hard glass, and 550 to 750°F for special glasses such as pyrex (most common). These elements produce powers between 1300 and 2500 watts for effective adjacent surface areas between 140 and 400 in^2

TABLE 2

APPROXIMATE EMISSIVITY OF VARIOUS SURFACES

Metals

Material	Polished	Rough	Oxidized
Aluminum	0.04	0.055	0.11-0.19
Brass	0.03	0.06-0.20	0.60
Copper	0.018-0.02	--	0.57
Gold	0.018-0.035	--	--
Steel	0.13-0.40	--	0.80-0.95
Lead	0.057-0.075	--	0.63
Nickel	0.045-0.087	--	0.37-0.48
Silver	0.02-0.035	--	--
Tin	0.04-0.065	--	--
Zinc	0.045-0.053	--	0.11
Galvanized Iron	0.228	--	0.76

Refractories, Building Materials, Miscellaneous

Material	Emissivity
Asbestos	0.93-0.96
Brick	0.76-0.93
Carbon	0.927-0.967
Glass, Smooth	0.937
Gypsum	0.90
Marble	0.931
Oak, Planed	0.895
Paper	0.924-0.944
Plater	0.91
Porcelain, Glazed	0.924
Quartz, Rough, Fused	0.932
Refractory Materials	0.65-0.91
Roofing Paper	0.91
Rubber	0.86-0.95
Water	0.95-0.963

Paints, Lacquers, Varnishes

Material	Emissivity
Black Lacquer	0.80-0.95
White Lacquer	0.80-0.95
Enamel (Any Color)	0.85-0.91
Oil Paints (Any Color)	0.92-0.96
Aluminum Paint	0.27-0.67
Varnish	0.89-0.93

(6.5 to 9.5 W/in^2), respectively. With pyrex emitting at 750°F, it is possible to obtain up to 13 W/in^2. For operating temperatures of these emitters, the emissivity of glass is 0.9 to 0.95, and therefore, close to those of a blackbody.

Vitrified ceramic radiating panels consist of a nickel-chromium resistance embedded in ceramic with a special ceramic enamel. The maximum permissible surface temperature for those elements is around 1300°F, but is normally between 750 and 1100°F. These emitters, which may be curved or flat, are available as rectangles or squares, for power ranges of 100 to 1000 watts (with surface areas varying from 7.5 to 23.5 in^2). Additionally, these elements come in a circular lamp configuration. Radiant efficiency of these panel heaters is low.

Other System Components

The two primary components of an IR system are the IR source and the reflectors. In addition, other components of an infrared system can include a cooling and/or ventilation system, a materials handling system and controls. Numerous types of IR sources have just been described. The reflector increases the system efficiency by redirecting the diffuse heating of the product. An additional purpose of reflectors is to protect other equipment and personnel from radiation. The efficiency of a reflector is dictated by its material, shape, and contour. Reflectors used for process-heating applications must not only be made of highly reflective material; they must also be able to maintain reflectivity over a long period of time, under the process conditions. The more efficient reflector (i.e., reflects most IR radiation, absorbs very little) will have a lower surface temperature, and therefore tend to stay cleaner, though all reflectors require periodic cleaning to maintain optimal efficiency.

Reflectors are usually constructed of steel, aluminum, or ceramic, and many also have coatings (or platings) of gold, silver or copper. The most efficient reflector coating is goldplating. A current ultra-thin gold plating, called Laser Gold (a proprietary process of Epner Technology, Inc.) has infrared reflectance of over 99%. A gold-plated coating significantly reduces the cooling requirements. Gold-plating on porcelain or extruded aluminum is commonly used for reflectors in very high temperature and high density applications. For lower temperature applications (below 600°F), aluminum is commonly used for reflector material because of its low cost.

One of the distinct features of infrared heating is the ability to control the direction of IR radiation by optically-designed reflectors so as to heat an entire product or only parts of the product. For example, when heating plastic sheets before forming, different areas can be heated to different temperatures to provide desired variations of flow characteristics of the material. Versatility and variations can be achieved using reflectors. In most cases, forced- or natural-movement air is required to assure satisfactory operating life for reflectors and electrical components (when present). In the extreme cases, water cooling is used. In higher-temperature applications, forced cooling of the reflector, source terminals and wiring is essential. A combination of shielding and forced convection can keep components which are temperature limited, such as the lamp and seals of quartz lamps, cool.

In addition, forced-air ventilation is required to remove water vapor, as well as to exhaust combustion gases in gas-fired IR systems and to exhaust toxic substances such as solvents (when present). The ventilation air in a gas-fired heater system serves the additional purposes of diluting the combustion exhaust gases to an acceptable temperature and providing supplementary convection heating to the product. For applications in which a solvent is being cured or dried, ventilation is required to remove the solvent from the work area for safety reasons. In some cases the ventilation system can also provide the cooling for the reflectors.

The material handling system or conveyor moves the product through the IR curing or drying phase. Many IR applications are retrofit to existing conveyors, and installation of only the emitter and reflector may be used to convert the system to IR. Some IR systems are modular units in which a belt transport system is incorporated into the unit.

A control system regulates the temperature of the IR source as well as the processing speed in order to ensure that the product is properly exposed. The control system can become very complex. The requirements of the application determine the selection of such items as the type of switching mechanisms or voltage control for electrical IR emitters (electro-mechanical, solid state contactor, phase shift, zero cross SCR, etc.), the temperature sensing device (thermocouple imbedded in heater or radiation pyrometer) and the type of control circuit (open loop, semi-open loop, or closed loop). Sensitive applications may require a microprocessor based system for controlling exposure time, distance from product and exposure intensity.

Performance

IR heating is a very efficient heating process, which results in high productivity, as well as low energy use for the finished product. Energy is also used efficiently due to the fast heating and cooldown capability. In addition, the product quality is very high due to the precise temperature controllability and clean oven conditions.

Efficiency, as it pertains to IR heating, can be measured by either the heating efficiency, which is a measure of the entire oven efficiency, or radiant efficiency, which is a measure of the heat source performance.

The heating efficiency of an IR heater is defined as the ratio of the theoretical heating power utilized to the heating input power. Utilized heating power is the heating power which must be applied to the material to achieve the desired temperature increase in a given time. The heating power required to vaporize water is included in the utilized heating power in drying processes. Typical losses result from ventilation, oven walls and cooling air or water. Heating efficiency is often used in oven evaluations and can be measured, though the value obtained represents an average value for the entire heating process. The heating efficiency of an IR oven largely depends on the following:

- Location of the IR heaters with respect to the surfaces of the material being heated (includes distance, direction and shape factors.
- Absorption coefficient of the material being heated and their reflection coefficients.
- Location of reflectors and oven walls.
- The ability of the emitter to generate the desired controlled pure wavelengths.

The radiant efficiency is the percentage of radiant output from a heat source versus conductive and convective output. As the temperature of a heat source is increased, the radiant efficiency also increases. The radiant efficiency is also dependent on the physical and ambient characteristics surrounding the heat source. Short-wavelength sources have radiant efficiencies ranging from about 65 to 85%, while long-wavelength sources have radiant efficiencies as low as 20%.

The combination of high heating and radiant efficiencies results in short process heating times. Product heat processing time cycles can be reduced from 1/5 to 1/10 of convection oven cycles.

Less energy is required to process a product by radiation versus convection because the energy can be directed onto the product, reducing heat losses to the air and the environment. In addition, IR ovens have very short heatup and cooldown times, so minimal energy is lost to heating the environment. This advantage also helps to decrease production time. Heatup times for electric IR systems can be a matter of seconds and for gas-fired IR systems, under 10 minutes, in contrast to convection ovens which usually require at least 30-minute warmup times.

Product quality is high as a result of accurate product temperature control and minimal contamination from air circulation. Because IR heat can be instantaneously controlled via radiation detectors, synchronous percentage timers and SCR's (Silicon Controlled Rectifier), the product temperature can be precisely controlled. This results in better finished products with fewer rejects. IR ovens are cleaner than convection ovens because air circulation can be minimized, thus reducing the quantity and circulation of contaminants, such as dust. Also, the faster cooldown rate prevents product damage from overheating. Electric IR systems are cleaner than gas-fired IR because no combustion products exist.

General characteristics of the most commonly used IR radiation sources are summarized. For most industrial applications, short-wavelength, tungsten-filament sources provide the high efficiency, deep penetration, and fast response rates needed. In contrast, long wave-length, nichrome filament sources have shallow penetration characteristics and are less efficient than tungsten sources, but are also less expensive and more rugged. Gas-fired IR heaters can be designed more economically for high thermal head jobs, particularly larger ones, than electric IR systems. But gas-fired IR heaters cannot generate the high temperatures (that electric IR sources can) required for efficient short-wavelength heating.

UV Lamps

Curing equipment most widely used has been ultraviolet lamps providing wavelengths of 300-400 nanometers (nm). It is estimated that 98% of all commercial photochemical radiation equipment in operation in the U.S.

uses ultraviolet lamps as their radiation source. Normal life expectancy is 2000 hr, or one year of one shift a day operation, according to the suppliers, but 1500 hr according to users. Use of lower-wavelength UV bulbs in a first curing of the surfacetric IR systems are cleaner than gas-fired IR because no combustion products layers, followed by a second curing using the 300-400 nm UV bulbs for the body of the polymer, is a technique for avoiding the need for inert gas blanketing.

The 200 W/in., medium-pressure, mercury-argon lamp with its 2000 hr working life and 63% efficiency (radiation output-power input) is the most popular unit today. Lengths are 1-7 ft. Because longer working lives and greater selectivity of wavelength cuts are desired, newer types of UV lamps are being introduced.

Newer Radiation Equipment

Electrodeless UV lamps, pulsed xenon and pulsed xenon-mercury lamps are being marketed for radiation curing. The electrodeless UV lamp (Fusion Systems Corporation, 11810 Parklawn Drive, Rockville, Maryland) is claimed to have a 50% greater operating life than UV lamps (3000 vs 2000 hr), to operate at lower temperatures and to operate at 29% lower line operating costs. In addition, Fusion Systems claims their bulb will cure UV coatings twice as fast and that one of their bulbs will replace 1.5-3.5 conventional 200 W/in. mercury vapor lamps. Fusion Systems also claims that a dryer with one 20 in. Fusion Systems radio frequency bulb would cost $14,600 compared to a conventional UV dryer with three lamps at $14,000.

Pulsed xenon lamps are intermediate power sources of UV radiation and have the potential for being the lowest cost lamps. In single lamp lots, conventional xenon lamps cost $125-$150 each, but in lots over 100, they can be purchased for under $100 each. The probable reason that pulsed xenon lamps are not used more widely is they simply are unknown to the UV coatings technologist at present. Only Autocoat, Inc., Middleton, Massachusetts, is known to use pulsed xenon lamps to UV cure links on plastic food containers.

Pulsed xenon lamp manufacturers seem to be ignorant of the technology of radiation inks and coatings. One potential problem with their use is the time interval of the pulse. Perhaps the coating speeds will be too fast to obtain uniform cure, even though the total power output during the pulsing period is much higher than for UV lamps.

Since pulsed xenon lamps have only now begun to be used to dry UV inks and coatings, it is simply too early to tell how much penetration they will make in these applications.

Safety

UV radiation below 220 nm and EB radiation equipment will produce ozone rapidly from atmospheric oxygen. As the radiation wavelength increases, less and less ozone will be generated. Current regulations by the U.S. Occupational Safety & Health Agency (OSHA) specify the time-weighted average concentration of ozone in the plant atmosphere over an 8-hr. working day to be 0.1 ppm (by volume). Proposed regulations would set the level of 0.05 ppm (measured at any time) as an "action level." Once the ambient ozone concentration reaches this action series of analytical measurements for ozone in various locations in the plant, the facility should institute a series of periodic medical examinations for his employees.

Because ambient ozone levels in many large U.S. cities are many times above OSHA's maximum allowable level of 0.1 ppm, it is believed that the action level of 0.05 ppm is meaningless. Health authorities may agree with the logic of this interpretation, but will counter that known health effects data indicate that exposure of human beings to levels of ozone above 0.05 ppm for lengthy periods of time does produce detrimental effects.

The user of UV and EB radiation equipment must, nevertheless, assure that the equipment he purchases will shield his workers from UV radiation (which can be accomplished by proper choice of materials and engineering design) and not produce excessive quantities of ozone.

The use of UV bulbs that produce sufficiently high wavelengths as not to produce ozone could be considered; however, radiation at these longer wavelengths is weaker, and therefore more UV bulbs must be installed per line length to accomplish UV cures in the same length of time. More bulbs will add to the operational cost.

Excessive ozone can be destroyed quickly by collecting the off-gases from UV dryers and passing them through an aqueous system containing a reducing agent, such as potassium iodide, sodium thiosulfate, or even activated carbon. Passage of dryer off-gases directly through activated carbon is not recommended, since heats of oxidation of carbon are exothermic and high. Should sufficient ozone be present, the carbon can become so hot as to start burning, and this could result in deflagration or even explosion, depending upon the quantities of ozone and carbon involved, rate of gas flow, degree of packing of the carbon and a number of other parameters.

Shorter wavelengths UV lamps can be used, but with a nitrogen blanket in the dryer. Ozone will not be produced with high-energy UV radiation passes through nitrogen. It has been found that this nitrogen blanket dramatically increases the drying rates of wet UV coatings and inks. Radiation energy formerly going to produce ozone now is used to cure the ink or coating. Additionally, once ozone is created, it will itself absorb UV radiation at a higher wavelength, thus sapping additional energy away from the curing of the coating or ink.

The benefits of eliminating ozone formation and obtaining much higher rates of cure (with less photoinitiation consumption) must be traded off against the cost of nitrogen gas.

Another method of controlling ozone is merely to vent it to the atmosphere, by means of the air fans drawing air into the dryer to cool the UV lamps. This is routine practice today, but as installations become larger, air pollution regulations become stricter, and more materials effects data quantifying the effects of atmospheric ozone upon plants and animals become known, catalytic destruction before discharge probably will be required.

APPLICATIONS

Coatings

Flat Wood Stock (Particleboard)

Europe consumes large quantities of UV coatings for flat wood stock. North American flat wood stock industry consume much less UV-cured wood coatings. The comparatively low U.S. consumption is most likely due to the fact that the European UV wood-coating technology is simply too slow for the scale of U.S. production facilities. Most European UV-cured wood coatings contain a small amount of incompatible wax, which exudes out of the wet coating to form an oxygen-impermeable barrier on the surface, preventing contact of air with the still wet coating. This wax takes time to exude out, and causes European curing times to be at least 30 seconds long. The U.S. wood stock industry, however, needs two-second photochemical cures to maintain productivity.

Conventional wet coatings are applied to particleboard panels and thermally dried at speeds of 80-150 ft/min. A nitrogen blanket will add costs to the process as will the use of EB curing, and the U.S. wood processing industry does not want to incur these extra costs on their production lines at this time. Infrared curing, however, is being studied closely by this industry because of lower costs for bulbs and because the coatings are conventional, and lower in cost.

For most wood coating lines, UV lamps are fast enough. EB accelerators and their accompanying radiation-shielding equipment cost 5-20 times more than the corresponding UV installation. Thus, the only potential for EB in the flat composition wood market is to cure at very high line speeds (30-40 m/min and up). At present, coatings are not available to cure at these speeds, thus the problem for EB in this market rests squarely upon the chemist to develop faster curing coating compositions. At the same time, new infrared thermal drying equipment is entering the market, making the future for EB curing in this application even less encouraging.

Containers, Closures and Metal Decorating

The market for containers, closures and metal decorating includes all finishes used to coat metal cans, crowns, closures and collapsible tubes. It also includes other specialty metal decorative finishes applied by lithography and silk-screen printing on various metal substrates, such as decorative nameplates.

Two technical innovations have affected major and minor can manufacturers in recent years all over the world. These are the growing use of "black plate steel" (which replaces tin plate on the basis of economics and must rely on organic coatings for chemical resistance) and the two-piece seamless can (replacing the three-piece can).

Three-piece cans generally are coated and printed before construction, that is, on flat stock-idel for radiation curing. Two-piece cans have what will become their interior surface coated with a heat-cured epoxy. The exterior surface is base coated, printed and top coated after construction. Uniform radiation curing of these exterior steps can be accomplished by rotating the can surface to receive the radiation.

Thus, targets for radiation curing are both the two- and three-piece metal cans.

In the U.S., however, there is an environmental movement that can market fear and may cripple the metal can industry. Some states already have passed legislation banning the sale of all disposable beer and soft-drink containers (metal or glass). Congress is moving to pass legislation requiring a 5-cent deposit on all glass and metal containers sold in federal government facilities. Should this environmental movement continue to gain momentum, the returnable glass bottle will likely gain in sales and reduce the markets won today by metal cans.

The inventor of the two-piece can, Coors Brewery (the fourth largest brewery in the U.S.) has ceased fighting the movement and instead has developed a machine for supermarkets and liquor stores that would crush returned cans into shippable bales and automatically return the deposit to the consumer. The energy required to recycle aluminum is much less than to produce the same quantity of aluminum from virgin bauxite ore. Coors has also developed and is preparing to test market a new all-plastic beer bottle.

Thus, although the can industry is one of the best potentials for radiation-cured coatings, the future of the metal can industry is in some doubt.

Radiation-cured coatings will not be used for interior can coatings, mainly because the ingredients of the coatings are too toxic, and no manufacturer wants to petition the Food and Drug Administration for approval in this application. However, exterior-based coats, inks and topcoat varnishes are all excellent markets for radiation-cured coatings.

With radiation curing, however, one liter of radiation-cured coating can replace 3.2 liters of conventional solvent-based coating, thus shrinking the total potential volume demand by 69%.

There are still some major technical deficiencies with UV-cured coatings for the metal substrate, can closure, crown, etc., markets. These are:

- Insufficient adhesion to metals.
- Poor flow-out and leveling characteristics.
- Marginal color stability when thermally baked.
- Marginal film flexibility and film toughness.

- Low pigment-loading capacity, resulting in marginal hiding properties.
- High coefficient of friction of cured film.
- Toxic, volatile acrylic monomers, thus making FDA approval for contact with food improbable.
- High cost: 100-150% more than conventional coatings.
- Radiation health hazards if UV dryers are misused.

Most of these deficiencies can be alleviated only by changing the chemistry of the radiation-curable polymer system; thus the future depends upon the chemist. Since the two-piece can-making system is replacing the three-piece process on the basis of economics, it would be ideal if all coatings and inks for the two-piece system could be radiation curable. But since the internal coating still must be heat cured (because of toxicity of radiation-cured compositions), the best technique that can be visualized until chemical breakthroughs are attained involves thermally baking the fully coated can at the very last step.

Motor Vehicles

American automobile manufacturers annually consume about forty seven million gallons of conventional waterborne and solvent-based coatings.

TABLE 3

ENERGY COMPARISON FOR THERMAL DRYING/UV CURING

In-Plant Ratios:	**Energy Savings**
Natural Gas Btu/UV Light Btu	86% fewer Btus
Oven Blowers/UV Blowers	93% fewer Btus
Oven Conveyors/UV Conveyors	75% fewer Btus
Gas In-Plant Electric Utility Original Source Ratios:	
Natural Gas Btu/UV Light Btu	68% fewer Btus
Oven Blowers/UV Blowers	93% fewer Btus
Oven Conveyors/UV Conveyors	

The Ford Motor Company was the only automotive company in the U.S. to use electron beam curing for many of its parts, and this operation consumed 100,000 gallons (0.2% of the total) in the fourth year after commercialization of this process by Ford. Since 1971, use of all EB-cured coatings in the U.S. has grown only at 4.7%/yr., far less than the growth rates for other new coatings, such as electrodeposition, waterborne primers and powder coatings. In the Ford operation, heat-sensitive plastic substrates are spray coated. The requirement for spray coating, in turn, necessitates the use of some organic solvent to reduce the viscosity of the coating sufficiently to allow spraying. The plastic parts being coated are truly three-dimensional in nature, and are not flat plates later shaped and formed.

All Printing Inks

UV-cured inks and overprint varnishes are used in two general market areas--packaging consumption and printing trade consumption. Folding paperboard cartons, used to package food, beverages, soaps and detergents, tobacco, cosmetics and medicines represent the second, prime, short-term packaging submarket for UV-cured inks and overprint varnishes. Printed labels are the third, prime, short-term submarket for UV-cured inks. Only on metal containers and web off-set paperboard cartons are heatset lithographic inks dried thermally. Because of the natural gas situation in North America, market penetration of UV-inks is virtually assured. It is estimated that 70% of all UV-cured inks will be used in these packaging markets. The remaining UV-cured inks will be sold to the publication and commercial printing markets.

Infrared Curing

Infrared radiation has been an important threat to UV curing, and its use has been pioneered in Italy. In fact, Italian wood spokesmen believe that UV-cured wood coatings have reached their penetration zenith, and that new wood coating lines will be cured with warm air and infrared lamps.

IR-cured coatings are strictly thermally cured. There are no photochemical reactions involved, only thermal evaporation of organic solvents from conventional solvent-based coatings and new, high-solids cadence, or water evaporation from conventional lattices and new waterborne coatings. Thermal energy reaches the wet substrate directly

TABLE 4

USE OF UV-CURED INKS AND OVERPRINT VARNISHES IN THE U.S.

Year	Sales Price ($/lb)	Packaging (metric tons)	Printing Trade (metric tons)	Total Value (million $)
1975	3.00	567	240.36	5.33
1980	2.17	6,136	2,630.00	41.95
1985	2.05	10,494	4,499.00	68.08
1990	2.05	15,274	6,549.00	99.31
1995	1.96	20,118	8,621.00	124.21
2000	1.96	24,907	10,685.00	153.92

TABLE 5

AVERAGE PROJECTED ANNUAL GROWTH OF UV INKS IN PRINTING (U.S.)

Period	Growth Rate (%/yr)
1975-1980	61.5
1975-1985	34.0
1980-1985	11.2
1980-1990	09.5
1975-2000	16.4

from the IR radiation, but also from metal reflectors. The IR filament itself produces operating temperatures of about 4000°F.

The by-product heated air is used to dry wet, coating films thicker than 2.5 mils. Hot air from the lamps is drawn to the front end of the IR line to a "flash-off" chamber. Here most solvents and/or water evaporate smoothly, prior to exposure of the coating directly to the IR radiation. In this manner, thick coatings can be dried without blistering.

IR radiation threatens UV radiation because of lower costs and because IR-curable coatings can be used in conjunction with proven, conventional coating technology.

IR bulbs available at $28-100 each compare with UV bulbs at $20-1000 each. Coatings for IR curing are conventional, selling for about half the price of UV coatings. IR electrical energy requirements are claimed to be about half those for UV, and are less than natural gas energy costs.

Conventional solvent-based and latex coatings, with their lower costs and wider spectrum of desirable properties can be IR cured. Use of conventional coatings eliminates the viscosity and toxicity problems associated with UV-cured coatings. Adhesion problems to metal are eliminated. In addition, IR radiation is oblivious to whether the film is clear, filled or pigmented. No shielding is required with IR dryers as with UV dryers.

The major limitation of IR curing is space for curing coatings thicker than 2.5 mils. To avoid blistering in these thicker coatings, the flash-off chamber is used, which must be at least 20 ft long (6 meters). For coatings of less than 2.5 mils thickness, only 1.0-1.5 sec IR radiation exposure are required to cure, in a 4-ft radiation chamber.

POTENTIAL APPLICATIONS

Pressure-Sensitive Adhesives

Pressure-sensitive adhesives is the single largest potential market for radiation-cured polymers. Photochemically cured adhesives could potentially replace polymers used worldwide to manufacture pressure-sensitive tapes and labels. Pressure-sensitive (PS) tapes and labels could cost less to manufacture by radiation curing. At the volume of polymers used, selling prices of well under $1.00/lb can be visualized for formulated EB-curable adhesives.

Future Economic Considerations

The major motivation for switching from conventional materials to radiation-cured formulations is the cost savings for natural gas which is used for thermally heating and curing conventional coatings and inks and for incineration after drying organic solvents. Additional advantages of radiation curing include smaller-sized overall production lines, increased production rates, lower total production costs, and the ability to coat or print heat-sensitive substrates. Many other additional process advantages accrue to radiation curing, but these depend on the specific application involved.

Various estimates have been made of the energy that can be saved by employing radiation curing materials in place of conventional, thermally cured coatings and inks. In coatings, Coors Brewery has estimated a potential 59 billion Btu in energy savings for the beer and soft drink can market.

The Can Manufacturers Institute estimates that if the metal and closures industry were required to incinerate all organic solvents from conventional coatings and if all then switched to UV or EB curing, the use of 11 bcf/ur of natural gas could be eliminated.

Cities Service has estimated total energy consumption for conventional manufacture of metal cans at 23.0 Btu per can, whereas the same can manufactured by UV curing requires 6.6 Btu per can.

Radiation Dynamics estimated that in the EB curing of polyethylene cable insulation, total costs (including amortized capital) for the conventionals extrusion plus steam process are 6.3 cents/lb vs. 2.5 cents/lb by EB curing. Of the 3.8 cents/lb differential, 0.8 cent/lb is in the higher cost for conventional insulation, and 3.0 cents/lb is the cost for peroxide (for the conventional formulation, but not needed in the EB cross-linked material).

Thermogenics, of New York, reported that a 38-in., 6-UV bulb dryer uses 50 kWh, the power cost of this UV dryer is $1.50/hr.

TEC Systems, Inc., a supplier of UV curing materials for the web off-set printing industry, estimates that for a 38-in wide web off-set press capable of printing speeds at 1000 ft/min, total operating costs of a conventional, heatset, high-velocity system are $50,327 vs. $33,532 for the UV curing system. These costs include installation and amoritized capital costs.

DeBell & Richardson estimated that for radiation curing of adhesives, to cure a 1-mil pigmented coating 48 in wide at the rate of 300 ft/min would require plant operating costs of $20.70/hr for thermal oven curing, $14.70/hr for UV curing and $7.26/hr for electron beam curing. These are plant operational costs only, and do not include capital costs. Included in these plant operational costs, however, are the following costs: thermal oven processing, requiring 5100 kWh at a power cost of $18.00/hr; or UV curing requiring 510 kWh at a power cost of

$10.20/hr; or EB curing requiring 40.2 kWh at a power cost of $0.81/hr.

Radiation Curing Potentials

Ultraviolet radiation-curing technology was introduced commercially in the United States in the late 1960s. Acceptance in terms of better product at lower costs had been gained prior to the natural gas shortage in the areas of lithographic printing inks, two-piece metal can coating and printing and fabrication of polymeric printing plates.

Electron beam radiation is commercial practice to produce heat-shrinkage polyethylene packaging film, to sterilize prepackaged bandages, for production of telephone wire, insulation of cables and coating of automobile parts. However, each practice is by a single company, and EB is not an industry-wide practice.

Major obstacles to both UV and EB curing are the rate of cure (faster speeds are desired) and improvement in properties of cured coatings, particularly with respect to metal adhesion. Formulations are desired that are non-toxic, allowing their use for food packaging.

Assuming that faster curing formulations are developed, which will allow faster production rates, the ability to coat the formulation prior to curing will become paramount. If the coating is too viscous to flow out and level at high production speeds, the advantages of increased rates of cure will have been lost.

The most promising applications for radiation-cured coatings are: (1) pressure-sensitive adhesives; (2) coatings; and (3) printed circuit boards.

TABLE 6

CHARACTERISTICS OF COMMERCIALLY USED INFRARED HEAT SOURCES

	Tungsten Filament Wire		Nichrome Spiral Winding		Low Temperature Panel Heaters		Gas Infrared Burners	
	Glass Bulb	T3 Quartz Lamp	Quartz Tube	Metal Sheath	Buried Nichrome	Metallic Salt	Perforated Tile Type	Impingement Type
Source Temperature Normal Max. Usual Range	4000°F 3000 to 4000°F	4000°F 3000 to 4000°F	1600°F 1800 to 1400°F	1200°F 1400 to 1000°F	600-800°F 1100-400°F		1700°F 1400-1550°F	2200°F 1400-2050°F
Brightness	Bright White Heat	Bright White Heat	Cherry Red	Dull Red	No Visible Light		Soft Red	Bright Red
Usual Size	G-30 Lamp	3/8" dia. Tube	3/8 or 5/8" dia. Tube	3/8 or 5/8" dia. Tube	Flat panels - Various		3 x 22 or 5 x 22"	3 x 12"
Wavelength at Energy Peak - Normal Max. (Emissivity)	1.15 Micron	1.15 Micron (0.86)	2.6 Micron (0.62)	3.1 Micron (0.56)	Around 4-5 Micron		2.6 Micron	2.2 Micron
Usual Range	1.5 to 1.15 Micron	1.5 to 1.15 Micron	2.6 to 2.8 Micron	2.8 to 3.6 Micron	3.2 to 6 Micron		2.5 to 2.8 Micron	2.2 to 2.8 Micron
Relative Energy Distribution Normal Maximum Radiation Convection & Cond. Usual Range Radiation Convection & Cond.	 86% 20% 65-80% 32-20%	 86% 14% 72-86% 28-14%	 55% 45% 55-45% 45-55%	 50% 50% 53-45% 47-55%	 40-30% 60-70% 50-20% 50-80%		 54% 46% 46-50% 54-50%	 44% 56% 36-40% 64-60%
Degree of Heat Penetration	Depth of penetration varies with the characteristics of the product. As a general rule, energy of shorter wavelengths penetrates deeper than energy of longer wavelengths.							

TABLE 6 (Continued)

CHARACTERISTICS OF COMMERCIALLY USED INFRARED HEAT SOURCES

	Tungsten Filament Wire		Nichrome Spiral Winding		Low Temperature Panel Heaters		Gas Infrared Burners	
	Glass Bulb	T3 Quartz Lamp	Quartz Tube	Metal Sheath	Buried Nichrome	Metallic Salt	Perforated Tile Type	Impingement Type
Relative Response to Heatup-Cooldown	seconds seconds	seconds seconds	minutes seconds	minutes minutes	scores of minutes scores of minutes		1 min to 60% 1 min to 18%	3 min to 60% 7 min to 18%
Color Sensitivity	Bodies of different colors can be heated at more nearly the same rate by infrared radiation with long wavelengths than they can be short wavelength infrared radiation.							
Ruggedness Mechanical Shock Thermal Shock	poor poor	good excellent	good excellent	excellent excellent	varies with panel design - could be quite good		fair excellent	poor fair
Average Life (hrs)		5000	10,000	10-20,000				

Source: Battelle Columbus Division, Radiation Curing: State-of-the-Art Assessment, June 1986.

3 PLASMA PROCESSING

INTRODUCTION

One of the most useful and efficient processes in the high temperature range is the plasma arc heater which heats gases by means of an electric arc. It goes far beyond the temperature range of conventional furnaces, extending from below 2000°F to almost any conceivably processing temperature, with efficiencies much higher than can be achieved with combustion heating equipment.

Applications for plasma systems exist in the chemical, metallurgical and specialty materials area. Applications for plasma systems include: blast furnace conversions, direct reduction or iron ore, hot blast superheating and hydrocarbon pyrolysis. Both the technical and economic aspects of these plasma applications will be described briefly.

Application of thermal plasma systems to the production of iron, steel ferroalloys and acetylene has been the subject of intensive research and engineering development over the past decade. The driving force behind this development effort is the escalating cost and decreasing availability of metallurgical coke and other fuels, which are used as fuel sources for both heat and reductant. Electricity prices are increasing at a much slower rate than those of the fluid fuels, and this trend is expected to continue for the foreseeable future. The steel industry over much of the world is further constrained by heavy commitment to aging in-place facilities, shortage of capital for their replacement and the need for strict compliance to environmental regulations. Plasma's attractions as a processing alternative lie in its independence of oxygen potential, flexibility in reductant selection, high energy density with resultant smaller process vessels and pollution control systems. The latter offers the strong potential for greatly decreased capital cost compared with existing technology.

Engineering developments have been undertaken to implement plasma for ironmaking in large scale pilot programs (~1000 kW) and the first production systems are undergoing initial operation. These applications include the installation of multimegawatt arc heaters for (1) injection of superheated reducing gas into blast furnace tuyeres, (2) generation of reformer gas for direct reduced iron pellet production, while (3) blast furnace wind superheating with plasma, and (4) plasma pyrolysis of hydrocarbons for acetylene production are presently in an experimental development stage.

Large scale pilot and production processes place a number of similar engineering requirements on the plasma system designer. Industrial systems are required to operate for over one week continuously on various process gases while providing high thermal efficiency and long term stability. Also, the plasma torch design has to provide the simultaneous capability to operate in a dirt ladened and often wet environment while maintaining electrical integrity and providing simple maintenance procedures.

A plasma is a state of matter formed by ionizing a gas, freeing electrons from their atoms. The resulting mixture of free electrons and positive ions makes plasmas good conductors of heat and electricity. Plasmas are produced by exposing one or several gases to a high-intensity electric arc, bringing the gas up to the plasma temperatures (as high as 20,000°F). The heat transfer properties and excellent controllability of plasmas make them efficient tools for industrial materials processing. The industrial application of plasma technology is still in its infancy. However, development has led to the availability of

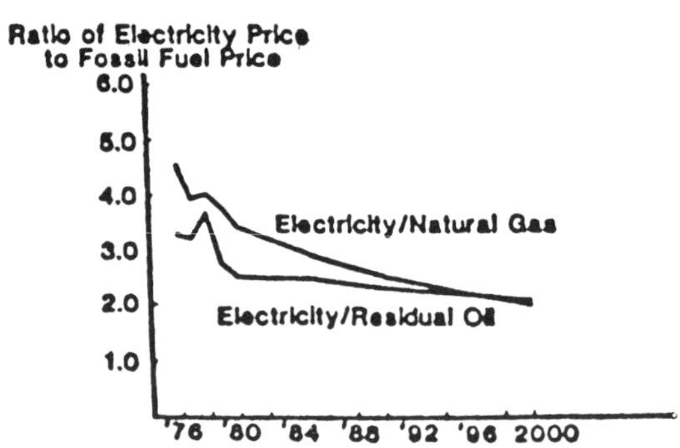

Figure 1. Energy costs for industrial users (USA).

reliable plasma torches in sizes up to 5 MW, a level that is adequate for most industrial applications of interest.

BACKGROUND

Matter has four states; solid, liquid, gas and plasma. When gas is superheated to temperatures about 8600°F, it is transferred from the gas to the plasma state. A plasma is a mixture of molecules, atoms, electrons and ions. Plasma is generated by exposing a gas to a high intensity electric arc maintained between two electrodes or by a rapidly changing electromagnetic field generated by induction, capacitive or microwave generator. Plasma generators can achieve temperatures as high as 20,000°F. This is far higher than the 2800°F practical limit for fossil fuel combustion. The device which holds the electrode(s) and creates the electric arc is known as a plasma torch.

Two types of plasma torches are used, transferred arc and non-transferred arc. In the transferred arc type, an arc is struck between the plasma torch and the material to be heated, with the torch acting as the cathode and the workpiece as the anode. An inert gas, usually argon, passing through the torch is the source of the plasma. In a non-transferred arc, both electrodes are in the torch. This type of torch can heat various gases such as argon, hydrogen, carbon monoxide or air to extremely high temperatures for chemical reactions or other processing. In the last decade, both transferred and non-transferred arc torches have been developed to the point where their reliability, electrode life and ease of servicing is sufficient for commercial operations.

Process technologies using plasma energy can vary widely, some with very different system configurations. At a minimum, however, plasma arc torches consist of a plasma gas supply, power supply, material handling system and control system for monitoring and regulating operation of the process. Plasma torches for heating gases normally have two electrodes. In non-transferred torches, both electrodes are mounted in the torch. One electrode is tubular and serves as the containment and path for the gas to be heated. In transferred torches only one electrode is in the torch, with the workpiece serving as the other. A schematic of transferred arc furnace is shown.

Industrial plasma heaters are designed and optimized for efficiency and geometry considerations unique to each process application.

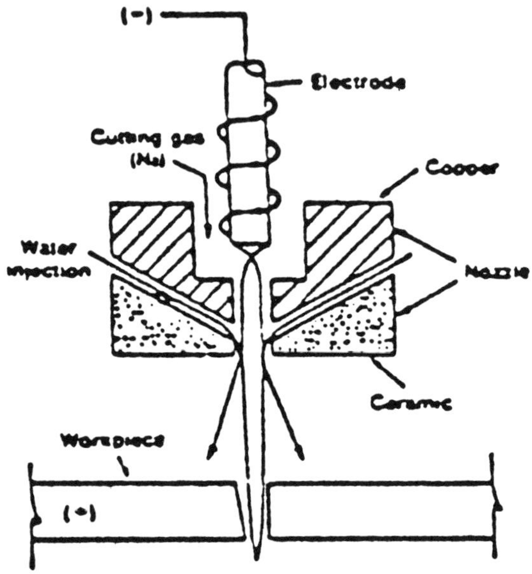

Figure 2. Transferred arc furnace.

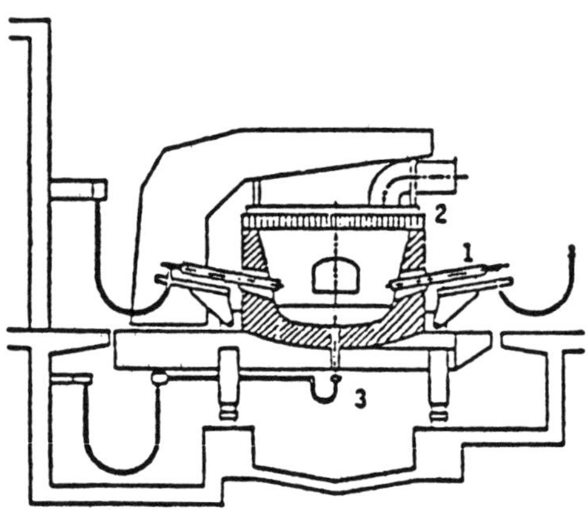

Figure 3. Non-transferred arc torch.

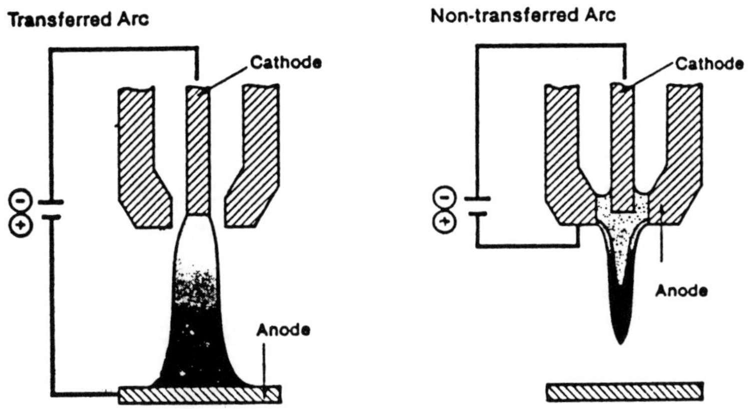

Figure 4. Types of plasma torches.

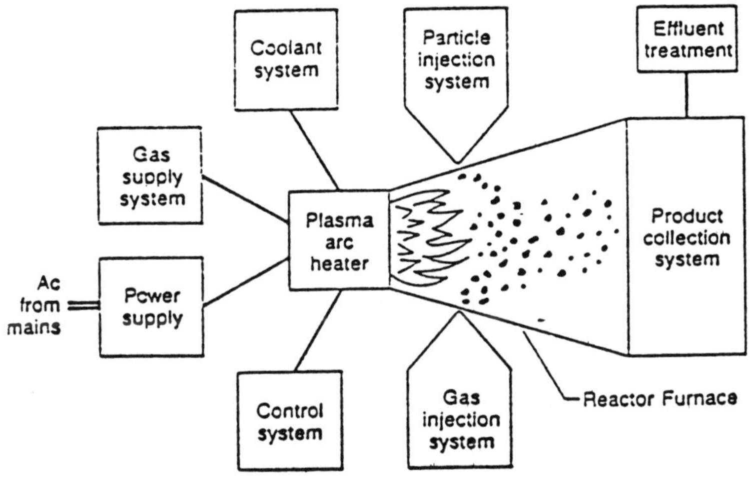

Figure 5. Typical plasma process system.

Depending on process requirements and the desired production rate, plasma heaters are specified in terms of power rating (wattage), gas requirement (type, pressure, temperature etc.), cooling water requirement, electrode material, and heater type and dimensions. Additional specifications may be required by the application. In the plasma cutting process, for example, plasma torches are also specified in terms of cutting rate capability (inches per minute).

Analysis of energy use is once again (as with many other electrotechnologies) misleading if no other factors are considered, e.g., "Plasmamelt" process developed by SKF in Sweden competes with the conventional blast furnace. While plasmamelt consumes 25% more energy than blast furnaces (17.5 million Btu versus 14 million Btu), it allows a shift to a lower grade (cheaper) coke, uses considerably less quantities of it and causes substantially less environmental problems. In other words, while energy consumption is higher for plasmamelt, costs are lower than those of blast furnace operation.

Costs of installation, operation and maintenance vary widely depending on the specific application. Some example applications include:

- Plasma Reduction: Plasmamelt process has several advantages including reduction in coke and maintenance requirements, ability to use lower grade (cheaper) coke and economically optimal plant size of only 100,000 tons/year in contrast to blast furnaces which are economic only at ten times the size. The annualized cost of a 250,000 ton/year plasmasmelt plant is $143.6/ton, while that of a 2 million ton/year blast furnace is approximately 25% higher at $178.20/ton. Half the reduction is attributable to reduced energy cost, while the remaining to reduced raw material cost.
- Plasma Melting: A conventional cupola furnace retrofitted with 6000 kW plasma torches costs approximately $4 million and is able to produce $94,300 tons/year of molten metal at an annualized cost of $41.70/ton. However, it saves $30/ton in reduced material costs through the use of cheaper scrap and $8/ton in energy costs and reduced silicon loss. The typical simple payback is less than two years.

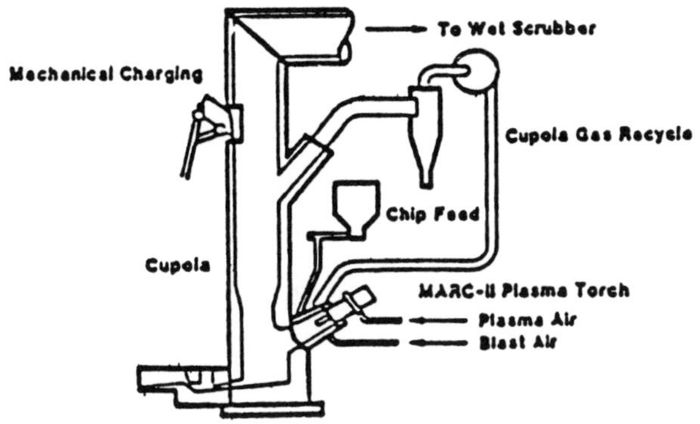

Figure 6. Plasma-fired cuploa.

- Plasma Cutting and Welding: Plasma cutting systems compete directly with oxfuel (oxyacetylene or oxyhydrogen) methods for cutting thick plates. Figures compares cutting costs of the two methods. The cost reduction in plasm torches is due to higher pressing speed, which is the predominant factor in selection of cutting and welding equipment. The use of plasma cutting and welding torches can result in 30 to 50 percent reduction in labor cost, 10 to 20 percent material saving (due to less scrap losses) and substantial increase in productivity (due to higher processing speeds). Additional material savings and higher production rates can result with the use of computer controls.

Plasmas are already well established in some metals fabrication processes notably welding, cutting and surface hardening. Several metals production plasma processes have also been recently commercialized. The availability of reliable industrial plasma generators gives plasma technology the potential for growth in several industrial sectors, most notably the production and fabrication of ferrous and nonferrous metals. A number of studies show great potential in processes such as steel ladle reheating plasma-baked treatment of stainless steelmaking

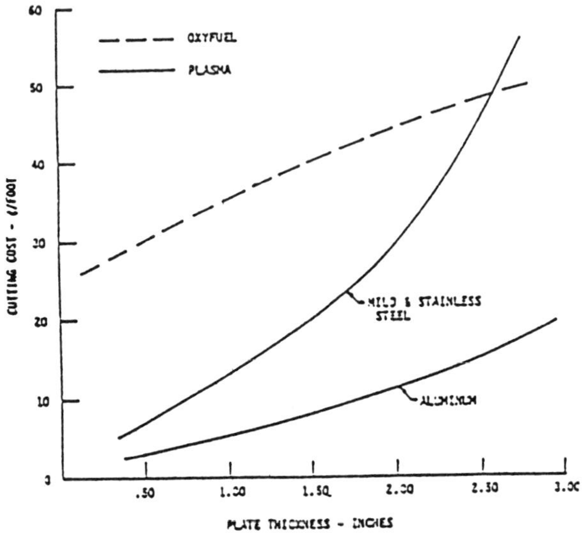

Figure 7. Plasma vs. oxyacetylene cutting.

furnace dusts, and the production of acetylene and ethylene. Plasma-based scrap melting technology is by far the most promising industrial plasma application.

Plasma applications illustrate many of the benefits that characterize electrotechnologies generally: electricity intensity, high temperature capability, outstanding controllability, increased energy productivity and more flexible fuel base compared with conventional combustion. In metals processing, the use of plasma may dramatically reduce requirements of costly metallurgical coke (thereby increasing capacity and lowering unit production costs), permit utilization of less expensive feedstocks (such as mill scale) and simplify manufacturing processes by eliminating the need for intermediate or finishing operations. Moreover, plasma heat is independent of oxygen potential and may be introduced using a wide variety of gases, thus permitting greater control over reducing atmospheres and product specifications. Reduced dependence on fossil fuels also may generate significant environmental benefits.

Among the industrial applications of plasma technology that are in use or being considered are:

Reduction Processes

- Direct reduction of iron ore to produce sponge iron.
- Smelting reduction of iron ore and scrap.
- Production of ferroalloys via plasma reduction.
- Recovery of metals from waste oxide dusts.

Heating Processes

- Boiler ignition.
- Destruction of toxic wastes.
- Fossil fuel combustion replacement in ore kilns.
- Metal melting.

Metal Fabrication Processes

- Surface glazing.
- Surface hardening of metals by tempering with plasma-generated heat.
- Plasma spraying a hardening material (e.g., tungsten) on a metal surface.
- Precision welding and cutting of difficult-to-machine metals.

The use of plasma energy for melting scrap iron, steel and alloys has progressed the furthest and is expected to be a major use of plasma processing in the U.S. Several commercial-scale scrap melting furnaces currently operate in Eastern Europe and Austria. One German firm has an AC plasma furnace in the latter stages of development, and a plasma-assisted foundry cupola has been developed in the U.S. by Westinghouse, Moden Equipment, General Motors and EPRI.

Plasma furnaces are not as economical as the electric arc furnace, but the potential for such systems is great. The advantages of plasma furnaces include the elimination of graphite electrode costs, higher temperatures and intensity, quieter operation, reduced air and water pollution, reduced alloy losses by oxidation and reduced arc flicker or power grid interference. The disadvantages are the short life time of the electrode, greater refractory wear due to long arcs and higher energy consumption due to heat losses caused by the long arcs. The major breakthrough needed is the development of a torch capable of high currents, such as 25 kA, resulting in shorter arcs.

PLASMA SYSTEM DESCRIPTION

A plasma heating system for industrial applications consists of five integral subsystems. These include (1) plasma torches for conversion of electrical energy to process heat by direct heat transfer between a rapidly rotating electric arc and a process gas stream, (2) a power supply with electrical characteristics designed to match those of the plasma torch, (3) a supply system to furnish the process gas to be heated, (4) a cooling water system for removal of the heat losses, and (5) a control and instrumentation system for interlocking the necessary subsystems and to provide performance data acquisition.

A plasma torch consists of a closely-spaced pair of tubular, water-cooled copper electrodes within which an electric arc discharge is magnetically rotated at extremely high speeds. In operation, a process gas is injected plasma through the gap between the electrodes which are spaced approximately 1 mm apart. A source of 4 kV electric power is sufficient to provide sparkover in the electrode gap, initiating the arc discharge, which is immediately blown to the interior of the arc chamber by the incoming process gas. There the arc is rotated at speeds up to 3000 revolutions per second by interaction of the arc current (up to 2000 amps) with a dc magnetic field set up by internally mounted solenoid coils. The patented sparkover feature provides a design which is self-starting when energized, and self-stabilizing during operation. Thus, spurious arc extinctions are completely eliminated. Since the internal surfaces are all water-cooled copper, virtually any type of process gas can be heated by the unit, whether oxidizing, reducing or inert. The combination of large throughput flow rates for the process gas and high arc rotation speeds produces a highly turbulent mixing action and yields a high operating efficiency (typically 75-90 percent) for equipment of this type. Process temperatures can be easily controlled by varying the arc current independently of the gas feedstock flow rate, and thereby optimized for high efficiency.

An example of plasma torch performance is shown in Figure 9. This data was gathered with the torch operating along a line of constant current. Some general features prevail about the characteristics of plasma torches. Efficiency and power increase with flow rate while gas enthalpy and temperature decline. For a wide range of gases, added gas enthalpies of 3 to 4 kW-hr/Nm^3 can be readily achieved at efficiencies

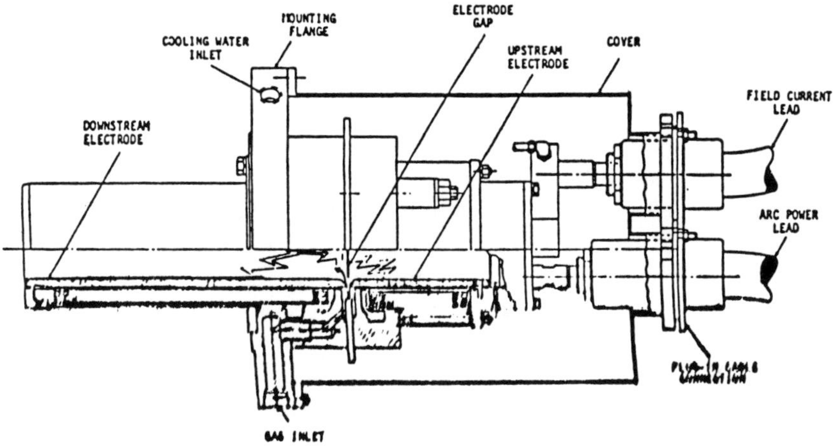

Figure 8. Cross section of plasma torch.

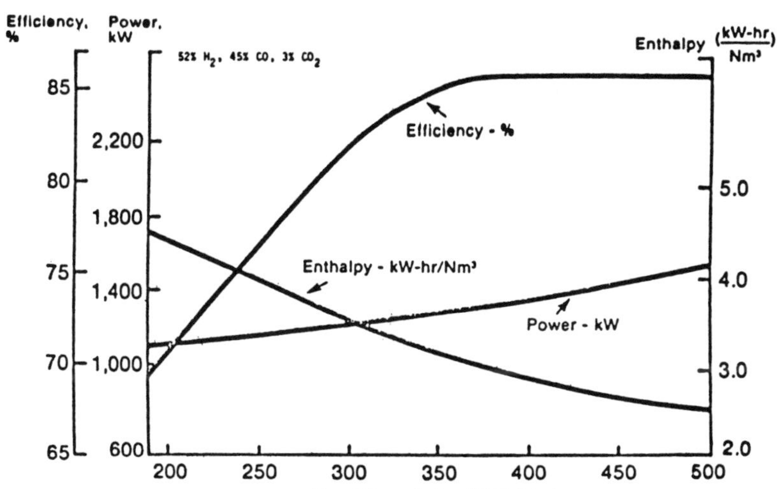

Figure 9. Plasma torch performance characteristics.

exceeding 80 percent. The gas temperature varies significantly due the variations in the specific heat of various gases.

Maintenance of these torches is mainly restricted to electrode changes, the frequency of which depends upon the gas flow rate, and power range required by the particular application. The time between electrode changes ranges from one to two weeks of continuous operation. Installation arrangements are designed such that electrode change can be performed in less than one hour. This is typically accomplished by separating the head from the base and replacing the downstream and upstream electrodes which are mounted on axially sliding O-ring seals.

Power supply characteristics must be closely matched to those of the plasma torch for optimum performance in an industrial environment. Plasma torches may be operated on ac or dc power; and power supplies must have sufficiently high voltage capability to provide sparkover in the interelectrode gap in order to assure process continuity. Electric arcs have strong negative impedance characteristics; i.e., arc voltage decreases with increasing arc current. The combination of strong electromagnetic and gas dynamic forces acting upon the arc column result in strongly dynamic variations in arc length, which result in high frequency variations in arc voltage.

Elements in the gas supply system will be dictated by the specific process requirements; utility sources or recycle streams may be used. In the latter case, dust laden streams must be scrubbed prior to compression thus providing a dry, particulate-free gas stream as the plasm torch gas source. If plan compressed air or steam sources are used, appropriate moisture separation and particle removal devices must be installed. Other gas system components include pressure, temperature and flow sensing, interlocks, flow control devices and appropriate isolation and shut-off valves.

Plasma installations for plant environments generally require closed loop coolant systems for removal of heat lost to the plasma torch walls. Local internal heat flux along the electrode arcing regions can reach levels of up to 2-5 kW/cm^2, thus high coolant velocities of about 10 to 30 m/sec are required. Circulating pumps with pressure of 15 to 30 bars total dynamic head are typically used. Other coolant system components include conventional valving, strainers, heat exchangers and surge reservoirs, as well as flow and temperature sensors for heat balance analysis and control purposes.

A typical control system for the plasma power supply consists of a console containing alarm indicators, panel meter readouts, pushbutton controls and associated relaying. The design of the control system includes an interlock scheme for start-up and shutdown which permits, under selected conditions, certain primary sensors to both lock-out and trip the power control breaker. Additional portions of the system include manual control of gas supply, coolant pump, field coil supply and capacitor and power transformer settings.

PLASMA-FIRED BLAST FURNACES FOR IRONMAKING

The injection of plasma superheated reducing gases into blast furnace tuyeres has been demonstrated to yield substantial reduction in coke rates. In pilot scale tests coke rates of 180 kg per ton of hot metal have been achieved, compared to the range of 400 to 700 kg per ton which is characteristic of modern conventional blast furnace practice. The potential for such large coke savings is particularly attractive as coke has become much less available due to coke plant obsolescence and declining capability. The problem for the steel industry in the U.S. is very serious, and a short-fall of 12 million ton (short) is projected, given a 2 percent annual growth in demand and a modest but predictable reduction in average coke rate to 560 kg per ton of hot metal.

Coke serves a number of functions in conventional practice blast furnaces. These functions include: (1) mechanical support for the burden within the blast furnace, (2) a porous medium through which blast furnace gases are distributed, (3) carbon reductant for oxides of iron and alloying elements such a silicon, (4) carbon source for iron carburization, and (5) a source of fuel for the blast furnace thermal requirements. It has been found in operating a pilot blast furnace of 0.3 m hearth diameter that the thermal requirement, in addition to the solution requirement, can be provided by superheated reducing gas, thus decreasing the coke consumption to that required for carburization of the hot metal, and reduction of the alloying elements. Coke rates down to 179 kg per ton of hot metal have been reported, which have been adequate to maintain furnace operation of the pilot unit.

A 3500 kW plasma torch has been installed and tested on a 500 ton per day blast furnace at Cockerill Steel in Seraing, Belgium, in order to determine its capability for operation in this type of service. This unit

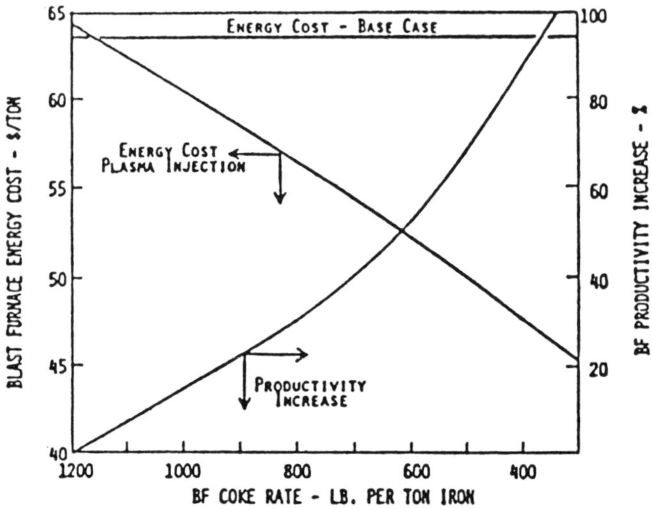

Figure 10. Blast furnace economics with plasma injection.

is totally containerized to function in the hostile blast furnace environment and is attached to the furnace inlet piping. Design characteristics of this large scale, self-stabilizing arc heater are:

- Arc Power - 1500 to 3500 kW.
- Gas Flow Rate - natural gas at 6000 to 800 Nm^3/hr.
- Thermal Efficiency - 75 to 80 percent.
- Blast Furnace Pressure - 1.5 to 2.0 bars.

Gas glow and energy requirements are designed to provide an overall air to gas volumetric ratio ranging from 2.4 to 3.2, entering the blast furnace at temperatures ranging from 2000 to 2400°C. Preliminary operation has indicated that a plasma torch is capable of reliable operation in the wet, dirt blast furnace environment. The tightly sealed, high voltage terminals and the container can be removed within minutes to provide access to the internal components for inspection or servicing. Electrical and thermal performance is within the designed range and the unit operates with good long term stability on both natural gas and natural gas-air mixtures; also thermal efficiencies are consistently about 80 percent.

BLAST FURNACE SUPERHEATING

Plasma superheating of the wind to the blast furnace provides a means to significantly reduce the coke required to produce hot metal. To plasma superheat wind, energy is transferred to the air stream via a plasma torch. To maximize the coke reduction due to wind superheating, other fuels such as oil, natural gas, coke oven gas and coal can be considered. To determine economic feasibility of trade-offs between coke and the alternative of fuels plus electricity, a blast furnace model was developed. The model indicates the fuels and electricity requirements with respect to specific coke reductions. Further, it is determined that increased productivity of blast furnace production rates can be obtained using plasma superheating of wind.

TABLE 1

EFFECTS OF PLASMA WIND SUPERHEATING FROM 1450°F to 2200°F

Fuel Input At Tuyere	Plasma (kWH/NTHM)	Reduction in Coke (lb/NTHM)	Fuel Added (lb/NTHM)	Production Rate of Nominal (%)	Fuel Cost* Savings ($/ NTHM)
Natural Gas	263	241	105	119	1.29**
Bunker "C" Fuel Oil	250	446	257	128	0.32
High Volatile "A" Bituminous Coal	246	538	440	132	21.71

Electricity Cost: 4¢/KWh
Coke Cost: $150/ton
High Vol. "A" Coal Cost: $40/ton
Nominal Production Rate: 100%
Natural Gas Cost: $3.50/MBtu
Oil Cost: 9¢/lb_m
Plasma System Efficiency = 80%

* Does not include productivity increase.
**Negative savings (i.e., loss).

PLASMA PYROLYSIS OF HYDROCARBONS

There has been interest in plasma-produced acetylene (from hydrogen or coal sources) as an alternative chemical feedstock to ethylene. The production of acetylene by arc processes offers a unique feature, in that the natural gas, hydrocarbon or coal requirement is utilized only as a raw material since the process energy is predominately electrical. Therefore, the price of acetylene is not totally dependent on the price, market conditions and availability of fossil feedstocks.

Methane is supplied to the arc heater from a pressurized tube trailer. The tube trailer capacity was about 55,000 standard cubic feet at a pressure of 2300 psig. The gas flow system consisted of a dome loaded orifice. Shut-off was provided by pneumatically-actuated ball values. Methane mass flow rates could be varied up to about 1.0 pound per second.

All arc heater operating parameters, i.e., arc current, voltage, power, water flow rates, inlet and outlet temperatures, feedstock flow rates, field coil current, etc., were simultaneously monitored using a 36-channel recording oscillograph.

The cracked gas stream was sampled from several locations along the 4.0 inch inside diameter exit pipe connected to the arc heater exhaust flange. Several water-cooled sample probes were used. As samples

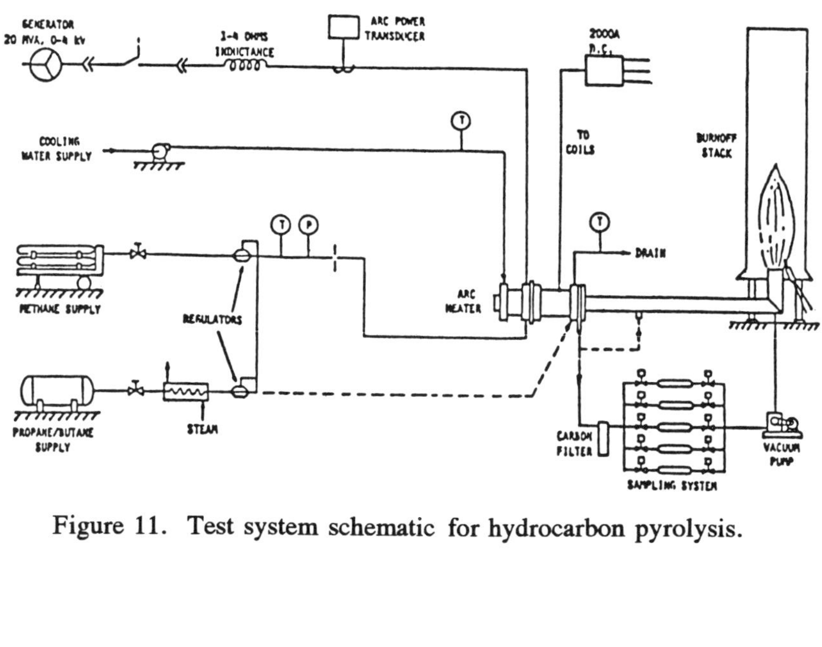

Figure 11. Test system schematic for hydrocarbon pyrolysis.

were removed from the stream, water quenched in the probes, flowed through soot removal filters to sample bottles, and exhausted through a vacuum pump. A manifold arrangement with solenoid valves was used to remotely collect multiple samples during individual tests. In this way, several arc heater operating parameters could be evaluated in a single test run. Analysis of the product samples was performed by a mass spectrometer.

PLASMAS

The use of electrical energy has long been an accepted practice in the chemical and metallurgical process industries. In many processes, particularly those of the metallurgical type, processing is characterized by high capital requirements, and growth in these industries has been limited. Thus, there is strong incentive to modify conventional processing methods to reduce both capital and operating costs.

During recent years new techniques have been developed, using electrical energy, which produced processing temperatures up to and beyond 5000°K. These temperatures are achieved in plasma media. One important characteristic of an electrically produced plasma is that energy can be supplied to a gas without loss of potential for chemical reaction. Energy supplied by combustion, for example, requires a substantial loss in reactivity since part of the gas is used to increase the temperature. With 100 percent reactive gases, extremely high temperatures, and rapid reaction rates, chemical, or metallurgical processing, in plasmas has significant potential.

Most of the current developments in plasmas processing were initiated in the 1960s when low cost electricity from nuclear power generating stations was forecast. This forecast has not developed. Nonetheless, the technological feasibility of this type of processing is sound, and the desire to reduce capital needs for conventional processing may continue to foster development work using plasmas for high temperature processing of raw materials to produce products of high need and value.

When most gases are heated at atmospheric pressure to about 5000°K, thermal ionization produces sufficient ions and free electrons to impart a fairly high electrical conductivity to the gas. As temperature increases, ionization increases to a point where the matter takes on the conductivity of metals. A plasma reactor, therefore, can be considered to be a device that contains high temperature ionized gas through which reactants flow and react to form products which leave the reactor, are quickly cooled, and are collected.

The simplest and oldest method of electrically heating a gas to plasma temperatures is the familiar one used in arc welding and the electric furnace. Here, the arc, or plasma, consists of ionized gas and electrode vapors in the region between the two electrodes and cannot be used independently of them. One advantage of the plasma produced by an electric arc is that a consumable anode may be prepared from graphite mixtures, and it can be fed into the plasma for participation in metal oxide reduction reactions. Feedrates of the anode can be adjusted to influence reduction rates. Another development to bench scale work, the radio frequency induced plasma, led to numerous investigations of chemical processes at plasma temperatures because the r.f. plasma provided a non-contaminating media in which the reactions could occur.

CORONA PHENOMENON AND SPECTRUM OF APPLICATIONS

Although plasma phenomena have been observed since antiquity, scientific inquiries were not made for many centuries, and industrial applications are rather recent. The premodern history of corona includes Benjamin Franklin's experiments in electrostatics around 1740, Faraday's discovery of electromagnetic induction in 1831, and the construction of the Wirmhurst generator in 1878.

After 1900, rapid progress was made in numerous industrial applications. Two key dates are 1906, when Frederick G. Cottrell invented the first electrostatic precipatator device, and 1935, when Chester Carlson started his search for a better copier. Table 2 identifies the key dates from antiquity to the present in the field of corona effect.

Corona is basically an electrostatic phenomenon. The corona discharge can also be said to be a plasma state.

Electrostatics is defined as ". . . that class of phenomena which is recognized by the presence of electrical charges, either stationary or moving, and the interaction of these charges, this interaction being solely by reason of the charges."

TABLE 2

HISTORY OF CORONA EFFECT

Date	Scientist	Finding
1600	William Gilbert	Published "De Magnete"
1740	Benjamin Franklin	Electrostatics
1831	Faraday	Electromagnetic induction
1906	Frederick G. Cottrell	Electrostatic precipitator (ESP)
1923	Detroit Edison Co.	First installed ESP
1935	Chester Carlson	Electrostatic copier
1944	Battelle Memorial Institute	Development of process eventually became Xerox

Plasmas are usually defined in terms of kinetic energy or temperature (°K) and particle density or pressure (particles/cm^2). The corona phenomenon (labeled glow discharges in Table 3) is a relatively low-temperature and low-pressure phenomenon.

In gaseous plasma, the ionization of gas by electron collision and by absorption of energy (photoionization) are the most important ionization processes. The electron emission from solids, especially electrodes, contributes greatly to the number of charged particles in a plasma. Also, when free electrons attach themselves to neutral atoms or molecules, negative ions are formed.

A plasma is technically defined as a neutral body of matter containing an appreciable concentration of mobile charge carriers (electrons and ions) in which the Debye length is smaller than any boundary surrounding the plasma.

Plasmas are classified according to pressure, charged-particle density, temperature, boundary conditions and the presence of external electric and/or magnetic fields.

Corona is a gas-phase phenomenon. Corona produces free radicals in a mixture of gases. In addition to reacting in the gas phase the radicals can also interact with molecules of a liquid or finely dispersed solid exposed to them.

TABLE 3

RANGES OF PARTICLE DENSITY AND TEMPERATURE VARIOUS TYPES OF PLASMAS

	Particle Density (No./cm^3)	Temperature (°K)
Natural Plasmas		
Stellar interiors	10^{22} -10^{25}	~10^8
Stellar atmospheres	10^{10} -10^{16}	10^4 -10^6
Nebulae	10^3	10^4
Interstellar space	1-100	10^2
Earth's ionosphere	10^{10} -10^{12}	10^2 -10^3
Manmade Plasmas		
Thermonuclear plasma	10^{10} -10^{14}	10^8 -10^9
Constricted arc plasma jets	10^{16} -10^{18}	1-5 x 10^4
Free-burning electric arcs	10^{16} -10^{17}	7-10 x 10^3
Combustion flames	10^{16} -10^{18}	3-5 x 10^3
Low-pressure arcs	10^{10} -10^{12}	1-3 x 10^3
Glow discharges	10^{10} -10^{12}	300-600

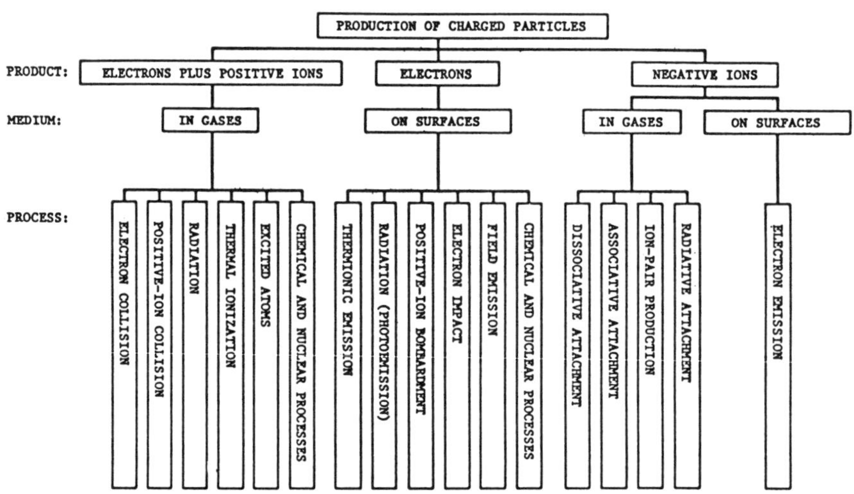

Figure 12. A schematic diagram showing the major possible processes for producing the various kinds of charged particles.

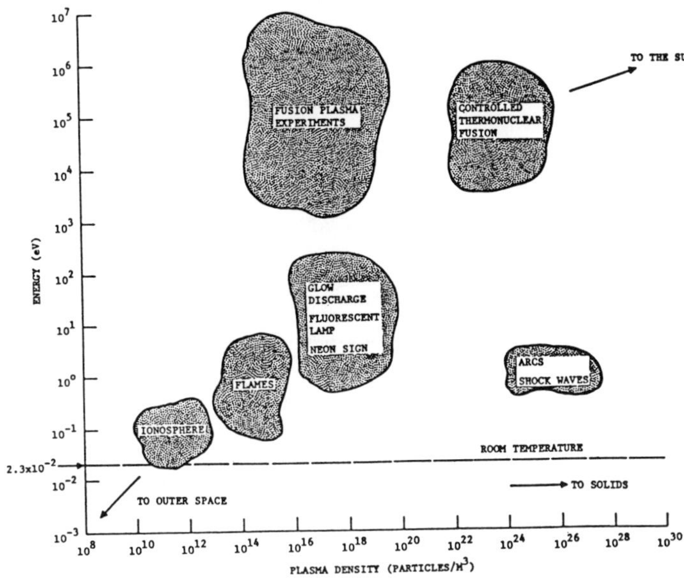

Figure 13. Energy-density diagram of the plasma state.

Shown in Figure 14 is a typical voltage-current plot for a positive needle-to-plane configuration in atmospheric air as driven by a dc power supply. Analysis of this curve and each of the distinctive regions allows descriptions of the phenomenon.

Region A. The phenomenon of corona starts with a few stray electrons always present in a gas because of the effect of cosmic and other background radiation. Originally, the electric field is low enough that the electrons do not have and cannot acquire the necessary energy to produce further ionization. If from an external source additional electron and/or ionization is introduced, the current will increase.

Region B. As the voltage is further increased, the current reaches a saturation level where all the charge externally generated will be swept out of the gap. Saturation curves will vary with the rate of external charge generation. This region is known as the Geiger regime.

Region C. At still higher voltages, the energy acquired by the electron becomes sufficient to ionize the gas by collision, forming additional electrons, leading to additional ionization. An avalanche occurs and the current rises rapidly. This is the Townsend avalanche

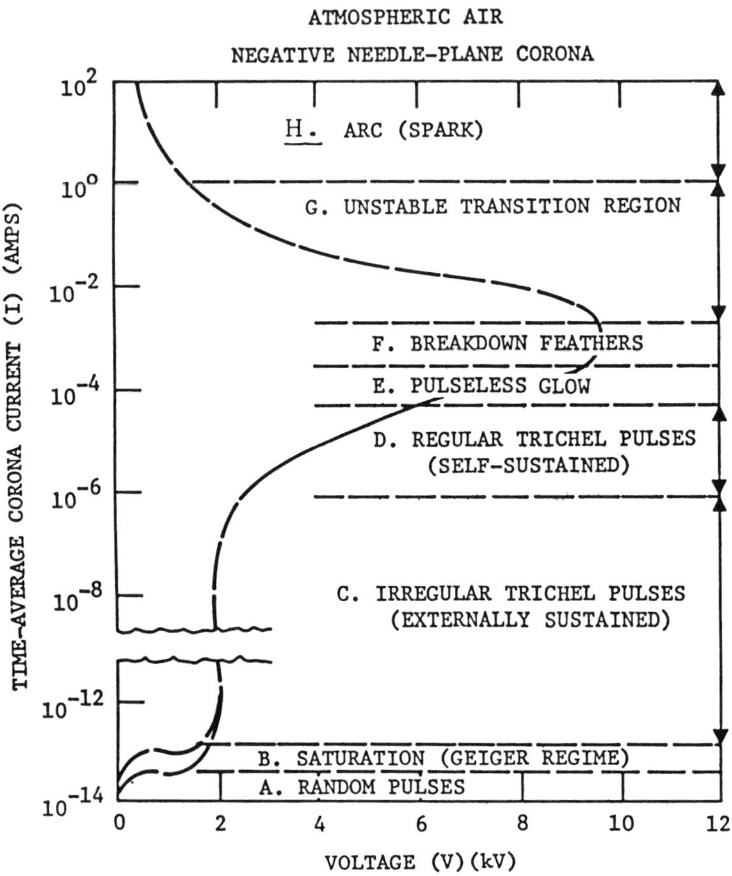

Figure 14. Schematic of current-voltage relation for air around atmospheric pressure.

region. Irregular current pulses, called onset streamers or burst pulses, occur.

Region D. As the voltage increases, these pulses become self-sustaining, i.e., no longer dependent on external ionization.

Region E. As the voltage is further increased, a steady state occurs as charges accumulate and continue the self-sustaining reaction. The anode starts to glow. This is the corona glow region. As electrons accelerate toward the positive electrode, they leave a trail of positive ions. Energetic photons ionize atoms and molecules in their path. Electrons following the channels of positive ions toward the cathode create an avalanche and the filamentary structure known as onset streamers. Eventually, the channels are dispersed due to charge repulsion and the positive ions drift away from the anode while the negative ions move toward the cathode.

Region F. At higher voltage, breakdown streamers appear. These are stronger than the onset streamers previously encountered.

Region G. When the voltage reaches a critical value, ionization occurs across the entire gap, creating an unstable transition region.

Region H. The unstable transition discharge evolves suddenly into a destructive high current arc or spark.

HIGH TEMPERATURE PLASMAS

This section deals with a limited set of industrial applications of high temperature plasmas. It shows how high-temperature plasma can play an important role in: (1) enabling new inorganic materials to be produced in a variety of shapes with desired properties; (2) fostering inorganic and organic reactions at rapid rates producing unusual (in the traditional chemistry sense) compounds; and (3) applying heat to a variety of materials. The most advanced applications have taken place in extractive metallurgy, but rapid advances are occurring in a variety of other areas.

In the field of industrial materials, the addition of heat at very high temperatures is necessary for the endothermic reaction, the heat of fusion, or a combination of both. At present, the source of heat is usually the combustion of gaseous or liquid hydrocarbon fuels. As the supply of these fuels becomes increasingly short and their prices rise accordingly, industry is beginning to look at alternate energy sources. Electrical energy, generated from the abundant supplies of coal and uranium, can be substituted for oil and gas firing. In many systems, electric arc heating would be more efficient than firing with hydrocarbon fuels because of the higher temperatures in excess of 2500°F (1370°C) is by electrical methods. The large fraction of energy that is normally discharged as waste heat could be considerably reduced when electric arc heating is utilized.

The field of plasma chemistry deals with chemical reactions in a partially ionized gas composed of ions, electrons and neutral species. This state of matter can be produced through very high temperatures or strong electric or magnetic fields.

Two regimes of significant interest in plasma chemistry are those labelled "glow discharges" and "arcs, jets or reactors." The first is characterized by average electron densities of 10^9-10^{12} cm^{-3}. An additional characteristic of such plasma is the lack of equilibrium between the electron temperature, T_e, and the gas temperature, T_g. Typical ratios for T_e/T_g lie in the range of 10-1000. The absence of thermal equilibrium makes it possible to obtain a plasma in which the gas temperature may be near ambient values at the same time that the electrons are sufficiently energetic to cause the rupture of molecular bonds.

By contrast, the conditions found in the plasmas produced by arcs, plasma jets or plasma reactors lead to an equilibrium situation in which the electron and gas temperatures are nearly identical. The very high gas temperatures ($>10^{3}$°K) measured in arcs and plasma jets makes them suitable for processing inorganic materials and organic compounds of very simple structures. Ordinarily complex organic materials and polymers cannot be treated under these conditions because they would rapidly be degraded due to their low thermal stability.

PHYSICAL DESCRIPTION OF THE PHENOMENA

A plasma jet is an arc-gas device that can generate extremely high temperatures; no combustion is involved. The plasma is the ionized gas created in an electric arc discharge in which electrons, positive and negative ions, and atoms are found. The discharge is characterized by intense luminosity. It is a good conductor of electricity and is affected by magnetic fields.

The electrical energy is converted within the plasma arc into other forms of energy - principally heat - and the heat is transferred by the mechanism of conduction, convection, radiation and diffusion. Conduction occurs by interparticle transfer from a region of higher temperature, called the source, to a region of lower temperature, called the sink. Radiation contributes to heating by the absorption and reradiation of photons by the medium. Convection depends on the difference in mass density of the heated gas and the main body of the surrounding gas. Diffusion depends on the concentration of molecules, atoms, ions, electrons and thermal gradients.

High-temperature physical chemical processing holds promise for the following broad areas:

- Highly endothermic reactions.
- Reactions limited at ordinary temperatures because of slow reaction rates.
- Reactions dependent on excited species.
- Reactions requiring high specific energy input without dilution by large volumes of combustion gases.
- Reactions or phase changes to alter the physical properties of a material.

Some of the advantages of thermal plasma processing include: rapid reaction rates, smaller apparatus, continuous rather than batch processing, automated control and useful new products. Thermal processing is made up of at least two large categories of phenomena; those of a purely physical nature and those involving one or more chemical reactions. Physical processing involves heat transfer between the plasma gas and the phase being treated, causing a substantial rise in temperature with associated physical transformation. Chemical processing involves one or more chemical reactions induced in the condensed phase or in the plasma itself.

The comparative efficiency of arc heaters versus a natural gas flame is shown in Figure 15. Assuming a theoretical flame temperature of 3560°F (1970°C) for the combustion of natural gas, the combustion products have an enthalpy of only 1300 Btu/lb (720 kcal/kg). Depending on the working temperature of the furnace, only a limited portion of the energy is made available. By comparison, the arc heater can heat air at 4000 Btu/lb (2200 kcal/kg) and make available a large portion of the energy over a wide range of temperatures.

CONFIGURATION AND DEVICES

Plasma arc devices are suitable for a variety of uses. The electric arc, which is constricted into a smaller circular cross section than would

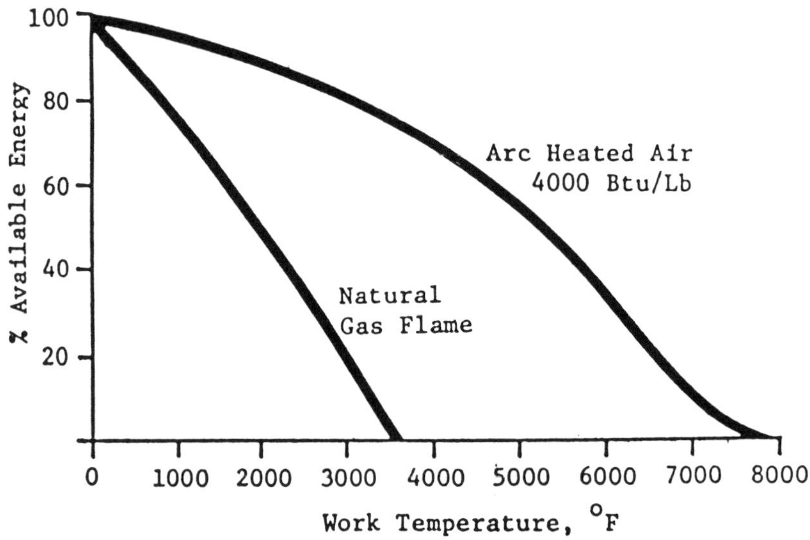

Figure 15. Efficiencies for high-temperature rating.

ordinarily exist in an open arc-type device, generates a very high temperature. This superheated-plasma working fluid can be channeled through an orifice and used as a reactive medium for chemical synthesis.

Plasma generators are classified as the non-transferred arc and the transferred arc. The difference between the two is related to the position of the electrodes with respect to each other and to the arc plume. A non-transferred arc consists usually of a cathode and an anode with an orifice or channel, so that when the arc is struck the arc plume emerges through this opening. The transferred-arc cathode is spaced some distance away from the anode and the arc is constricted between both electrodes.

Although many plasma jets have been perfected for various purposes, the most common type used for chemical processes is a direct-current, gas-stabilized plasma arc of the type shown in Figure 16. The reactor chamber may be of any configuration needed to accommodate different feeding and quenching devices. A schematic of a plasma reactor is shown in Figure 17. Gases or powders are injected into the plasma from a ring attached to the bottom of the plasma generator. The chamber can be fitted with quench tubes of various shapes to cool the products.

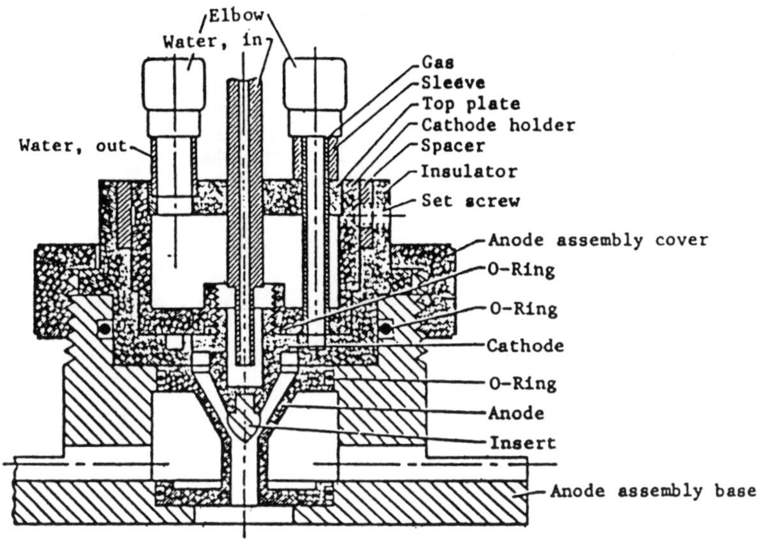

Figure 16. Typical plasma jet.

There are two other ways to characterize plasma devices: one uses an arc between two electrodes to generate the plasma; the other does not employ any electrodes, and electrical power is coupled through in induction coil.

Plasma generators are described in the literature either as open arc or the constricted arc of the plasma jet. The stabilized arc plasma generator is the only unit currently able to achieve ultra high temperatures for extended periods.

Finally, plasmas can also be classified by the method for providing the plasma arc. Radio frequency induction coupling can be used in electrodeless generators. The common dc arc torch can incorporate both consumable and nonconsumable electrodes.

Plasma Gas

The choice of plasma gas is very important. The gas can be part of the reaction or (more often) serve as an inert carrier. Noble gases like argon and helium which have low ionization energy and good arc voltage, can serve as carriers.

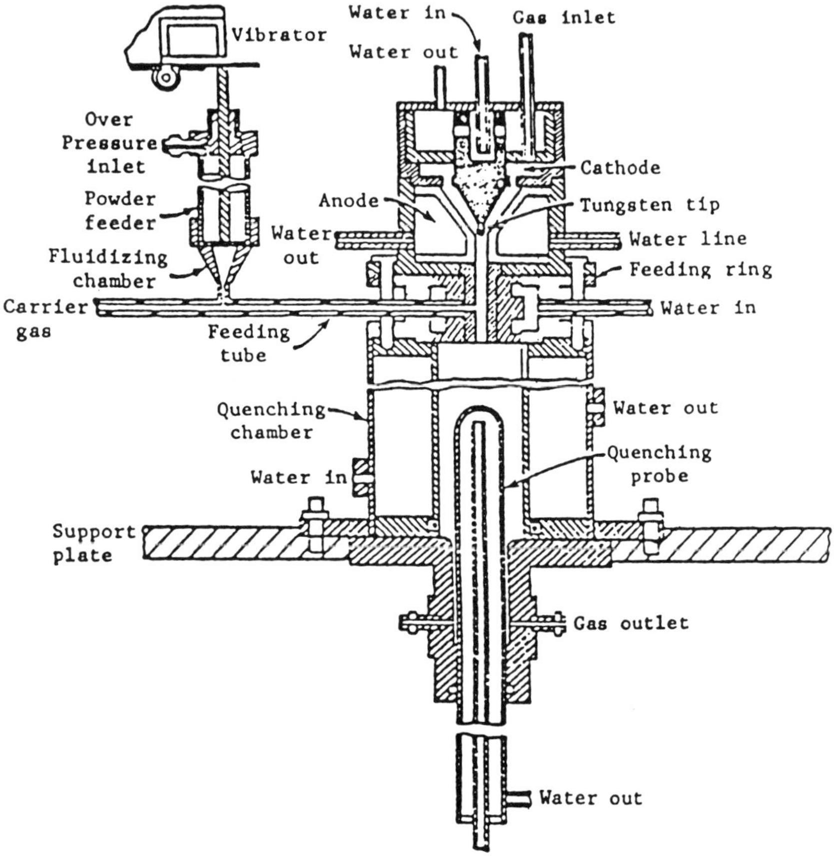

Figure 17. Typical plasma reactor.

Electrodes

Graphite, tungsten and copper are the three basic materials commonly used as electrodes in plasma torches. While a graphite electrode requires no cooling, it is consumed in the plasma generation process and, therefore, electrode feeding devices are required. Also, the abated graphite electrodes are a source of plasma stream contamination that can be reduced to negligible levels in certain types of plasma furnaces.

The design of a plasma jet can be varied to meet the requirements of the chemical process. An example is the introduction of a reactant at a certain point along the flame path. Consumable cathodes have been used

TABLE 4

PROPERTIES OF GASES USED IN PLASMA APPLICATIONS

Gas	Dissocia-tion Energy (kcal/g mole)	Particle after Dissociation	Ionization Voltage (V)	Open Arc Voltage[a]	Arc Temperature[b] (°F)	Heat Content of Gas[b] (W/ft³) (NTP)
A	0	A	15.68	18	18,000	75
He	0	He	24.46	26	27,000	110
H_2	104	H	13.53	70	15,000	260
O_2	110	O	13.55	40	16,000	425
N_2	225	N	14.48	40	16,000	425
Air	--	--	--	60	16,000	425

[a]Constant arc length and power.
[b]10% thermal ionization.

in experiments in which carbon was one of the reactants. Many experiments use carbon vaporized from a graphite cathode in the chemical studies. The introduction of a powder carried in a gas stream or as a constituent of a gas either admixed or premixed has also been used to admit the reactant into the plasma stream. Tungsten or 2% thoriated tungsten electrodes are the most frequently used water-cooled, non-consumable cathodes. Water-cooled copper anodes have been widely used in many arc generators designed for chemical synthesis.

Quenching

Cooling the desired products of a useful reaction at high temperatures, in a limited time, to give a desired yield and selectively with minimum energy waste is the major problem associated with wide-scale industrial use of thermal processing. A variety of quenching-methods are in use.

Surface Heat Transfer

Conventional surface heat exchange is often adequate for cooling high-temperature gases. Water-cooled copper walls have been used for quenching argon plasma. The maximum heat flux through a fixed surface is limited to about 500 Btu/sec/ft^2 (575 W/cm^2). The heat flux of some plasmas has been reported at 10 times this amount, limiting the areas of applications.

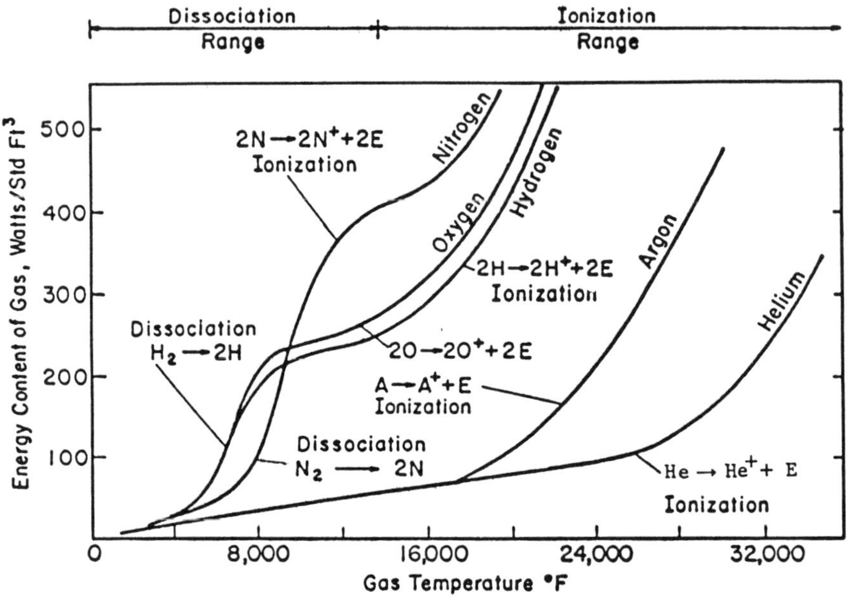

Figure 18. Plasma temperature as a function of gas energy content at atmospheric

Use of Secondary Components

Nonreactive media, such as inert gases or relatively inert solids, can be brought in direct contact with the high-temperature products. The fluidized bed method is especially advantageous for quenching solids. Temperature decay curves (quench rate) of the order of 50 x 10^6F (28 x 10^6C)/sec have been calculated.

Expansion Techniques

Expansion through a Laval nozzle, intermittent expansion into a ballistic piston and expansion through a turbine having transpiration cooled blades have all been proposed. Quenching rates of the order of 30 x 10^6F/sec (17 x 10^6C/sec) are predicted.

TABLE 5

RANGE OF PLASMA APPLICATIONS

Plasma Interactions With	Range of Applications	Equipment Used
Compact Solid Phase	Welding Cutting Spraying Drilling of rock Building up matrix material Charge preheating Lining curling	Burner cutter
Compact Liquid Phase	Melting Remelting Refining Alloying Reduction smelting Crystal productions	Furnace
Disperse Condensed Phase	Ore beneficiation Metal reduction Inorganic synthesis Organic synthesis Refractory material processing Powder processing Spraying	Plasma jet reactors

Power Supply Sources

Thyristors with associated arc current stabilization are the most often used power supply sources for dc plasma generators. Performance of 1000 V/1000 amps are common; power sources of up to 7 MV-amp available. For small plasma generators, silicon-diode power supply sources rated at 350 V/600 amps are available.

For high-frequency plasma, energy generators and transformers with a 10-12 kV constant anode voltage assembled on thyistors or semiconductor diodes have high efficiency and practically unlimited power. High-frequency generators often use electrovacuum parts.

FUTURE PROJECTS

A major improvement in plasma technology will be the development of generators that can operate on oxdizing substances such as air or steam. Electrodeless plasma generators also have demonstrated considerable advantages over those with consumable electrodes. New generator designs in the making will probably take advantage of magnetic pinch, gas stabilization and geometry to better control heat flux delivered at the materials surface; ac power for large voltages will replace dc because of better economics.

Continuous countercurrent contact operations will replace batch systems. Improvements in quenching should take place rapidly and additional new techniques to enable the separation of the desired products should appear.

4 LASERS

INTRODUCTION

Laser is the acronym for *Light Amplification by Stimulated Emission of Radiation*. The phenomenon is shown in Figure 1. By imposition of an external source of energy, such as a voltage field or a flash of light, an electron may jump from its normal energy state to an "excited" state of higher energy. When the excitation is removed, the electron will decay to its stable state, emitting a packet of light, or photon. If a lasing medium is repeatedly excited, a population inversion of the normal energy states can be forced to occur and a large fraction of the atoms in the materials can be plated in the excited state simultaneously. As a few of the electrons in the medium change their normal states, the photons they emit stimulate the transition of other electrons, and a cascade effect occurs, generating an intense light pulse. The typical lifetime of the inversion-decay cycle is on the order of a nanosecond (one-billionth of a second), so, in fact, the excitation can be applied at such high frequency that it may be considered continuous for practical purposes. In addition to the source of excitation and the lasing medium, a means of optical feedback must be provided. This amounts to a pair of mirrors placed at the two ends of the optical cavity, one of which is totally reflecting back and forth within the optical cavity, further stimulating photon emission, while 5% or so is allowed to "leak" out to provide a usable energy source.

Laser light has characteristics which make it particularly valuable in industrial applications. It is monochromatic, i.e., of fixed wavelength; its wavelength can be selected by proper choice of the lasing medium and, to some extent, adjusted by design of the optical cavity to match specific applications. More importantly, the light is coherent, i.e., the photons emitted by stimulated emission in the lasing medium are all in phase. This characteristic has important implications for the way laser

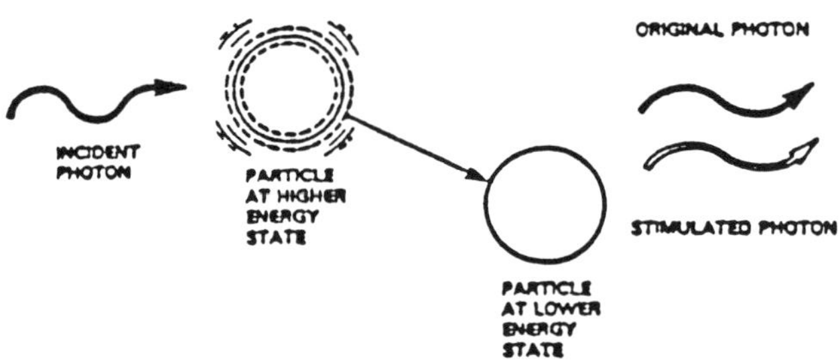

Figure 1. Principle of stimulated emission.

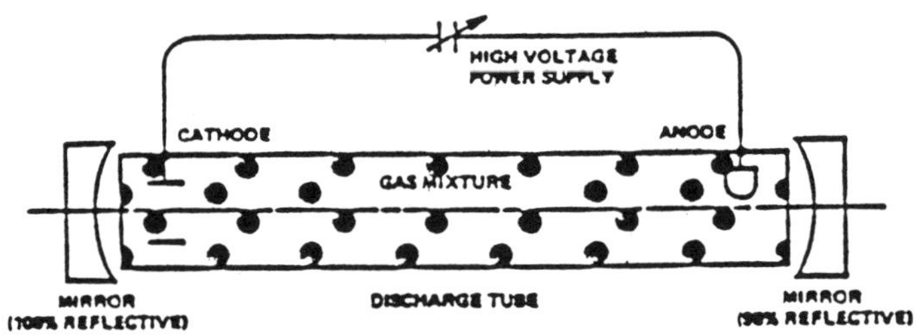

Figure 2. Laser optical cavity.

light interacts with materials. Lasers produce an exceptionally intense light source that can be optically focused with great precision, concentrating the energy. This ability to concentrate the energy is significant in many applications where the intensity of energy deposition per unit area is more critical than the total power contained in the beam. Several types of lasers are currently in use for industrial applications.

Conversion of electrical energy to thermal energy deposited in the material is an inefficient process in a laser, ranging from only 1% to 10% in efficiency. This disadvantage over conventional methods is offset because laser energy can be very precisely concentrated and thus, affects only a fraction of the material processed by conventional methods. Therefore, when comparing the relative energy requirement to carry out a given process, the product of energy input per volume times the volume of material affected, laser processing has advantages.

However, energy may be insignificant in other processes such as cutting. Here the quality of the product, i.e., precision of the cut and the production rate are of greater importance. In such a case, lasers can be more advantageous.

APPLICATIONS

Laser applications are easiest to justify in cases of high production volume given that lasers have high production rates and best cost-effectiveness at capacity. The high production volume may involve a wide range of industries and products - from tobacco to semi-conductors.

The high power density and precise controllability achievable with lasers open a broad variety of opportunities for their use in materials processing. These include applications in cutting, welding, surface treatment and scribing of both metallic and non-metallic materials. The

TABLE 1

COMMON LASING SYSTEMS IN MATERIALS PROCESSING

Media	Excitation	Resulting Radiation
Solid State: Ruby (Chromium in Aluminum Oxide)	Photon Absorption	Visible Red Light (06943 Micron)
Neoymium (ND) (In Glass or in a Crystla of Uttrium Aluminum Garnet, YAG)	Photon Absorption	Infrared Light
Gaseous: Carbon Dioxide (In Helium and Nitrogen Mixture)	Collision	Infrared Light (10.6 Microns)

nature of the application dictates the required specific energy (joules/cm^2), which, in turn, is the product of the power density (watts/cm^2) and the interaction time (seconds). Shock hardening induces highly localized thermal stresses on the material surface by impinging a beam of very high power density, but for such a short pulse that melting does not occur. At the other end of the spectrum, transformation hardening requires longer, less intense pulses.

In cutting and drilling, precise focusing of the beam produces intense, localized energy deposition, resulting in vaporization of the material. For cutting, the required laser power is dependent on the desired cutting speed and the thickness and physical characteristics of the material. In many applications, relatively low power (500 watt or less) lasers are suitable, even for cutting metals, when coupled to an optical system that focuses the beam to a spot a few thousandths of an inch diameter. Laser cutting has been applied to such diverse applications as the cutting of fabric in the apparel industry, stripping of plastic insulation from wire, cutting of titanium sheet for aerospace applications, and the cutting of fragile ceramic substrates for microelectric circuits. Systems for these applications are always numerically controlled to permit easy adaption to complex geometries. With computer control, the intensity of the beam can be varied as needed for the given application, and the material can be moved under the focused spot. Control systems are also available that position the beam either by moving the laser head or with mirrors. These systems permit relatively rapid motion, but in some cases, pose difficulties with defocusing.

Lasers have been widely used in applications requiring rapid, precise drilling of small holes, such as the metering holes in flow controllers for washing machines. Larger holes requiring precise positioning and orientation can also be produced with lasers; transpiration cooling holes in gas turbine blades made of difficult-to-machine, high-temperature alloys are one example. In applications where multiple holes are needed, the beam can be optically split; non-circular holes can also be drilled by optically shaping the incident spot.

Welding applications of laser range from spot welding of miniature parts, such as relay contacts, to full-penetration welds in steel, titanium, and superalloy plates up to 1/2" thick. Lasers provide a non-contact weld with precise depth control. The application of lasers in material processing is a young industry, characterized by rapid technical and

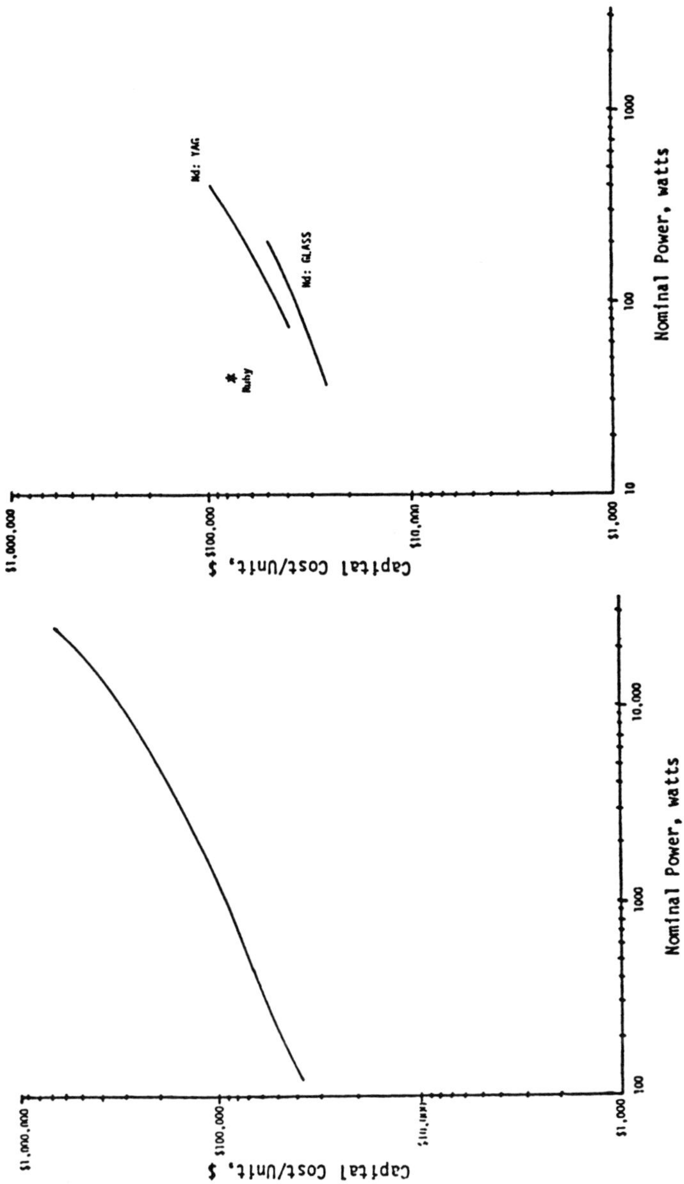

Capital Cost of CO_2 Lasers vs. Power

Capital Cost of Solid-State Lasers vs. Power

Source: "Assessment of Materials-Processing Lasers," Report #EM-3465, Project #1967-3, EPRI, May 1984.

Figure 3.

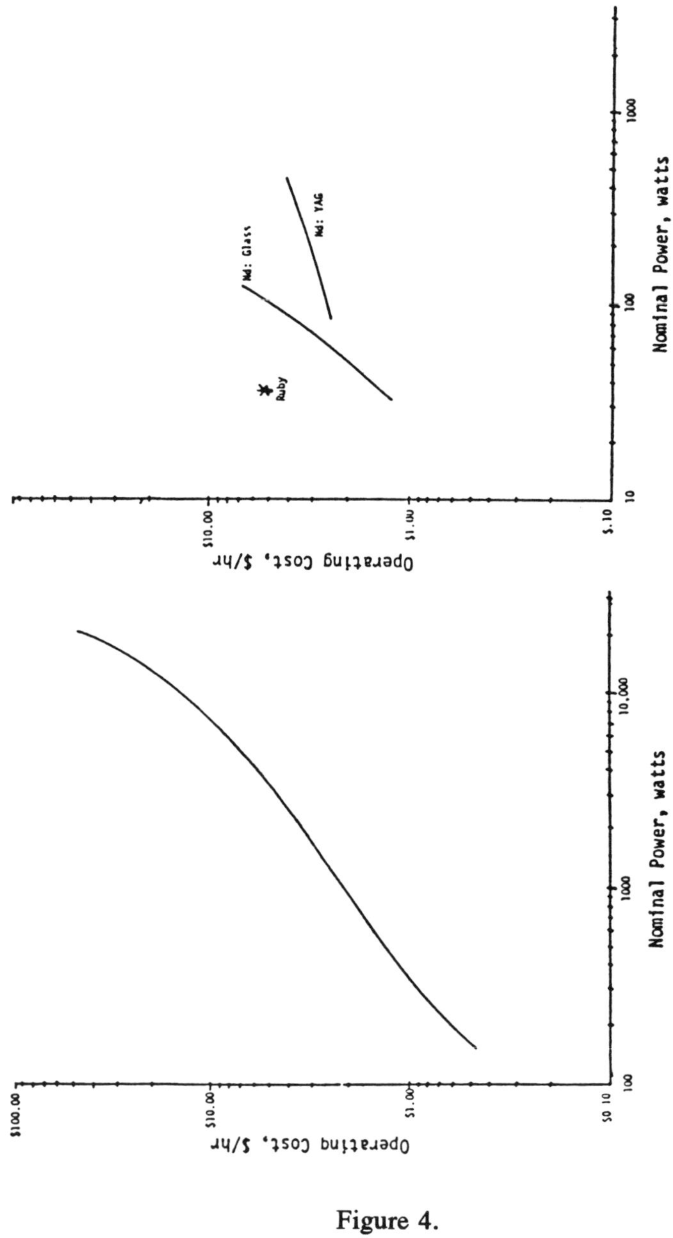

CO_2 Operating Cost vs. Power

Solid-State Operating Cost vs. Power

Source: "Assessment of Materials-Processing Lasers," Report #EM-3465, Project #1967-3, EPRI, May 1984.

Figure 4.

market growth. There are about 35 manufacturers in the U.S. of laser material processing systems and components.

TABLE 2

APPLICATIONS OF LASERS

Manufacturing Category	Example of Usage Reported
Cigarettes	Removal of filter
Men's & women's suits	Cutting of cloth
Cardboard boxes	Cutting of board
Pharmaceutical products	NA
Plastic hose	Cutting of tubing, welding of metal attachments
Fabricated rubber products	Cutting rubber
Products made of glass	Cutting of glass, welding of metal
Ordnance & accessories	Welding of launch supports
Turbines	Removal of combustion liners
Farm machinery & equipment	Removal of plate holes
Food products machinery	Welding of machinery
Pumps & pumping equipment	Welding of equipment
Telephone equipment	Welding of equipment
Computer peripherals	Welding of equipment
Heart assist devices	Precision cutting
Aerospace	Welding, heat treating
Aircraft	Welding, heat treating
Pens, mechanical pencils & parts	Cutting, welding
Heat exchanger	NA
Air conditioning (auto)	Welding of housing
Watches, clocks & parts	Precision cutting
Ophthalmic goods	Precision cutting
Electric motors	Welding of parts
Appliances	Welding of parts
Semiconductors	Precision heat treating, resistor trimming
Hybrid electronics	Resistors trimming
Storage batteries	Welding
Truck & bus bodies	Welding of parts
Motor vehicle parts & accessories	Hardening steel parts
Railroad equipment	Heat treating cylinder liners
Surgical appliances & supplies	Welding batteries
Dental equipment	Welding braces

PURIFICATION OF MATERIALS

The high selectivity of lasers enables the photodissociation of impurity molecules without affecting the desired component. This has been investigated for the purification of materials for semiconductor use.

It has shown that minor amounts of 1,2-dichloroethane, $C_2H_4Cl_2$, and carbon tetrachloride, CCl_4, can be decomposed and removed from arsenic trichloride reported the purification of silane (SiH_4). Investigated has been preferential decomposition of phosphine (PH_3), arsine (ASH_3), and diborane (B_2H_6) in the presence of silane. In mixtures containing 100 parts of silane to one part of impurity they have shown that with an argonfluorine (ArF) laser greater than 99 percent of the arsine was removed while destroying only one percent of silane. Removal of phosphine and diborane was also achieved, greater than 40 percent removal with destructions of six and two percent of the silane, respectively. Later studies showed that thc dcgrcc of purification and its efficiency increased when the temperature was reduced from 295°K to 198°K. Laser purification has promise as an economical process for silane purification.

MICROELECTRONICS FABRICATION

The use of lasers in microelectronics has recently been discussed. This use of lasers takes advantage of its high coherence with resulting localization of the induced reaction rather than on frequency selection. Consider this to be an exciting field, still in its infancy and with high potential applicability in microelectronics fabrication at the large scale and very large scale integration levels. Laser chemical vapor deposition is being investigated for depositing thin films of a variety of materials on semi-conductor or insulator substrates.

Studies on the decomposition of metal from alkyls, such as dimethyl cadmium (DMCd) and trimethylaluminum (TMAl) have been made. Irradiation with a 250 nm laser has been shown to sequentially break the bonds in DMCd leading to vapor phase formation of metal atoms followed by their deposition. The details of the bond breaking steps from TMAl have not been as well resolved. Optical micrographs of the resulting deposits show very good spatial resolution. Based on the data, anticipated is an ultimate resolution of the deposition to be finer than

0.5 μm. The rate of metal deposition has been shown to be linear in laser energy flux (fluence) and the partial pressure of DMCd. Cadmium deposition rates of greater than 100 nm per second have been observed. This process enables both localization of deposition and surface heating.

Laser photoetching has been demonstrated by the photolysis of an methyl halide with a laser beam. The halogen atom formed can react with the semiconductor to produce an etch pattern on the surface. Visable etch marks have been formed on gallium arsenide surfaces by this procedure.

Recent work on laser annealing, originally emphasizes the annealing of defects created by ion implantation, has been extended to transient heating of semiconductors. Involved is the understanding of fast crystalline growth involving both the solid and liquid phases. Problems associated with laser annealing are under investigation.

The following conditions must be met for the successful application of a specific approach to isotope photoseparation:

- There must be one absorption line in the species being separated which does not overlap any absorption lines of other molecules in the mixture.
- The required monochromatic radiation must be available with the necessary power, duration, divergence and monochromaticity.
- The primary photophysical or photochemical process must enable the easy separation of the excited species from the mixture.
- The selectivity of the absorbing species must be maintained throughout the separation process.

The following properties of lasers which make them an excellent tool for meeting the above separation criteria:

- Tunability of radiation frequency enabling covering the frequency range from infrared through vacuum ultraviolet.
- Controlled duration of the radiation to less than the lifetime of the excited states of the irradiated species.
- Spatial coherence which enables forming directed beams of radiation with long path lengths.

- Monochromaticity and temporal coherence allows the selective excitation of a given energy state which is only very slightly different than those of other energy states of the species present.

A large number of isotope separations have been explored experimentally. Here the status of the separation of deuterium from hydrogen and ^{235}U from ^{238}U will be briefly reviewed. These two cases have been selected not only because of the differences in the mass ratios of the isotopes involved but also because these are the isotopes of greatest commercial interest.

Currently, the most active investigations for deuterium separation are based on:

- Multiphoton dissociation processes of deuterium containing organic compounds using pulsed carbon dioxide lasers.
- Infrared induced bimolecular reactions using CO or CO_2 lasers.
- Single photon dissociation with UV laser.

The first of these methods has led to an enrichment of deuterium exceeding 10,000 Freon 123, and 3,3-dichloro-1,1,1 trifluoroethane, using multiphoton dissociation, have been most extensively investigated. The deuterium containing freon, CF_3CCl_2D, preferentially dissociates to $CF_3{=}CFD$. With a laser wavelength of 10.65 μ and a freon pressure of 30-100 Torr, deuterium enrichments of 1200 $\pm$ 300 have been reported. A higher enrichment factor, 11,000 $\pm$ 2000 has been observed using difluoromethane as the starting material.

Biomolecular reactions investigated have included the reaction of hydrogen halide and water:

$$DX^* + H_2O \rightarrow HDO + HX$$

*excited

and the two-step photodissociation of ammonia

$$2NH_2D \rightarrow 2HD + H_2 + N_2$$

The UV laser photolysis of formaldehyde shows promise as a single-step method for deuterium enrichment UV absorption which can lead to formaldehyde decomposition via a predissociation mechanism. This involves a light frequency greater than that required for dissociation which excites the molecule. The redistribution of the energy leads to dissociation. Enrichment factors of 900-fold have been achieved.

The laser approaches to enrich U are purely physical or a combination of physical and chemical methods. The uranium-metal laser approach is based on the vaporization of uranium at 2000°C, the selective two-step ionization of the desired isotope, and separation of the ionized metal from the remaining vapor by means of an external electric or magnetic field. In the combined physical and chemical process, uranium hexafluoride is expanded by means of a nozzle to form an essentially collision-free molecular beam. The UF is selectively excited to the first vibrational state with an infrared laser and then exposed to a UV laser which creates an excited electronic state. The electronically excited UF can form solid UF or a reactive uranium-fluorine species which can be scavenged. Another beam method is based on the divergence of UF molecules due to the additional momentum resulting from their absorption of laser energy.

An interesting photochemical approach to the laser separation of ^{235}U is based on the isotope-selective photodissociation of the volatile uranyl hexafluoroacetylacetonate tetrahydrofuran. This has been photo-dissociated in a molecular beam using both a continuous wave and a pulsed carbon dioxide laser. The O-U-O stretching frequency, which is a function of the masses of both the uranium and oxygen isotopes involved in the bond, was selectively photolyzed. The isotope selectivities observed were close to theoretical. The extent of reaction was directly proportional to the laser fluence.

The status of the development of uranium isotope separation technology is uncertain. Jersey-Nuclear-Avco-Isotopes were reported in *Laser Focus* (February 1980, pp. 20-22) to have applied for permission to build a privately funded U recovery plant in Washington state. This was later changed to a request for support from the Department of Energy, and construction was to begin in early 1981. The process selected for demonstration was the one based on the laser ionized uranium metal vapor beam.

MISCELLANEOUS LASER APPLICATIONS

Several other studies of laser chemistry have been directed toward specific applications. Improved conversion of 7-dyhydrocholesterol (7-DHC), to previtamin D by a two-step laser photolysis. The two-step photolysis reduces or eliminates competing photoreactions. The conversion of 7-DHC to previtamin D is high, greater than 90 percent, and the extent of contamination is small.

Production sinterable powders of silicon, silicon nitride and silicon carbide by the laser photolysis of silane, silane and ammonia, and silane and ethylene, respectively. A CO_2 laser was used with the reference experiments using an energy level of 760 W/cm^2. The reactant pressure was 0.2 atm, and the flow of the reactants was at the right angles to the laser beam. The powders formed were spherical with diameters of 12 to 100 nm with a standard deviation of 25-45 percent. The purity of the product was very high as was the efficiency of the process. Approximately 95 percent of the silicon hydride reacted in a single pass through the laser beam. Approximately 2 kWh of energy were required per kilo of silicon nitride formed. Both the silicon nitride prepared by nitriding silicon and the laser synthesized silicon nitride were converted to densified silicon nitride bodies. The small particle sizes led to rapid sintering.

LASER PROCESSING OF MATERIALS

Metals can be sheared, slit, punched, drilled, notched, nibbled, cut and sawed to produce a variety of parts for further assembly or finished products. A very complex edifice of science and technology based on fundamental studies and accumulated experience has been erected over the years. While the methods and techniques are well established and time tested, a great deal of room for improvement exists.

Lasers are being applied to a variety of industrial manufacturing tasks: the welding of parts; the heat treating of surfaces to improve properties; the cutting of metal, wood, cloth and plastic parts; and the drilling of ceramics and rubies. Lasers serve basically to apply high flux energy to a small area of the surface of a workpiece.

A number of advantages are associated with such a system, including a product of higher quality, reduced material losses, higher productivity, a treatment in surface causing less damage to the workpiece (stress, strain), a more acceptable working environment and greater flexibility and versatility.

A number of industrial processes, using light, are reviewed.

DRILLING

Drilling and deburring is an energy expensive process and lasers do not compare well with traditional methods except in special cases. Drilling by laser is favored in the case of small holes, when high precision is required or if the material is difficult to drill by mechanical means. Glass lasers using pulse duration of from 150 to 1500 microseconds are commonly used for drilling. This system is especially applicable if a large number of very small holes (0.007" to 0.040" in diameter), drilled at acute angels in very hard material are required. Great success has been achieved in drilling rubies used in watchmaking, diamonds used as dies for drawing wire, ceramic substrates in the electronic industries, transcription holes in turbine blades, bleeder holes in cold rolled steel, fuel pump valves and very hard materials used in the aerospace industry. The lasers have been very useful in making very small holes in a variety of materials: polyethylene, plastics, etc. Typical drilling speeds of commercial systems are 1200 holes/minute, but this rate can be increased to 10,000 per minute in some applications.

CUTTING

The major advantage of laser in cutting is the ability to follow, under computer program, a very complex pattern, providing high precision, repeatability, flexibility and productivity. Major applications include cutting preformed metal sheets, organic fibrous materials, fiber reinforced plastics, ceramic material, quartz, glass composite materials and fabric.

Medium power carbon dioxide lasers provide low operating costs and high quality product in sheet metal operations. A sharp, clean edge at high cutting rate and with minimum set-up time is observed.

Some of the major advantages of laser metal cutting include:

- Minimum material waste.
- Minimum set-up time.
- No wear of the cutting tool.
- A smooth, clean edge can be obtained at high cutting speed, not requiring further cleanup operation.
- Little distortion or damage to the workpiece from heat input. The same applies in terms of mechanical stress or damage.
- Sharp contours profiles or complex shapes can be easily executed.
- Hardened materials can be cut easily.

Some interesting applications of laser cutting include cutouts in painted panels for meters, gauges, louvers, etc.; trimming of parts; duct workpiece parts from a continuous roll of galvanized steel; petrochemical seal rings cut from sheet stock; custom designed fuel tanks; and stainless steel signs with stylized letter cutouts.

With advances in using a wider variety of metals and alloys, lasers have been used to cut stainless steel, beryllium, magnesium alloy and tungsten. Lasers have also been used to cut thin non-metal materials, quartz, aluminum and epoxy material.

WELDING

Laser beams produce good crystal structure with improved mechanical properties of hardness, tensile strength and impact strength. The fact that no welding rods, fluxes or protective material are required is a plus. The superb advantage of laser beams in welding is their ability to reach difficult to access places and to be fully automated with high precision to handle complex shapes. The laser has found a home in the industry requiring the welding of dissimilar materials.

Some advantages of laser welding include:

- Very localized heating.
- Ability to join two dissimilar metals.
- Metallurgical control.
- Real time verification of the weld.

SURFACE TREATMENT

A number of metal surface treatments are ideally achieved using lasers.

TRANSFORMATION HARDENING

Heat treating of a surface without the addition of new material causes the material to change from one solid phase to another. The heat applied only in surface, is rapidly quenched by the bulk cold material. A layer a few millimeters thick is sufficient to give the material hardness and resistance to fatigue. In the case of laser glazing, the surface layer becomes totally amorphous or glassy.

Lasers produce slightly higher hardness than other techniques, due to more rapid quenching. Lasers can be used to perform more selective hardening.

In order for laser energy to be absorbed by reflective metals, the surface must be treated with a coating that absorbs the laser energy. This affords another opportunity for selective hardening by using different masking patterns.

CLADDING

This is a process adding to a base material, a layer of another material, to give the entire piece desirable properties. In strip cladding, a coating material in the form of a ribbon is continuously applied to a surface and fused to the base material with a laser.

Cladding has advantages over other coating processes in that metallurgical mixing occurs at the cladding interface. Laser cladding is preferred since it produces less dilution (mixing of the base metal with the clad metal) than traditional techniques.

ALLOYING

Materials in the form of sheets, rods, powders or rings are added to the base material by the action of heat. The laser melts the material which diffuses into the base to a predefined depth (controlled by the operating conditions). The rapid cooling gives the alloyed layer a fine microstructure.

The main advantage here, besides giving the piece unusual properties, is to save on rare or expensive material, since the material of interest is produced only in a thin surface layer. The advantage of laser is the ability to treat only a precise area.

MELTING

Laser surface melting can be used to produce amorphous metals or metals with a glassy structure. These metals are highly resistant to corrosion. Laser can remelt a surface to improve wear resistance.

The output beam from a laser must be shaped before it is applied to a surface. The most common beam shaping technique is focusing with a transmitting lens or a reflecting mirror. Sharply focused beams can be directed to any point on a surface and are usually used for melting larger diameter; focal points are used in conjunction with scanning, beam integration and variable beam shaping techniques.

MACHINING

Photomachining is derived from photoengraving. A photo negative of the master drawing is prepared and used to contact paint the component images onto a metal sheet covered with a photo sensitive coating. The developed image acts as a mask to the etching solution used to dissolve unprotected metal areas. The method permits the handling of very intricate designs and is used mostly for prototype or short run jobs.

WIRE STRIPPING

A number of applications require stripping wires from their insulation material. Thermal methods are judged to be too slow for production type operations, while mechanical stripping requires frequent calibration because of tool wear.

Using a CO_2 continuous laser, concentrated energy is focused on the rotating wire, rapidly melting and vaporizing the insulation without affecting the wire or its plating material. Oxygen introduced in the system serves as gas-jet assist, oxidizing the insulation. A vacuum system removes debris and vapors, insuring a safe system. The high speed system requires very little set-up time.

PRODUCT MARKING

Marking information is desirable on a number of products for identification purposes, to convey product information and for theft prevention. Laser marking can be achieved by (1) engraving: microdrilling a groove in the workpiece by vaporization of the material; (2) dot matrix marking: vaporizing a series of matrix marking; vaporizing a series of minute holes (typically 75 micron in diameter, 4000 holes/second); (3) mask imaging marking: the beam illuminates a mark which acts as an object for a projecting lens.

Laser marking is costly but finds applications when:

- Small characters are required - characters as small as 12 microns in width can be obtained.
- A permanent marking is required.
- The piece cannot be steel-stamped because of its shape, size, fragility, hardness or precision.
- Marking must be done without contacting the piece.
- High speed is required.
- Great positioning accuracy for the marking is needed.
- Chemical contamination from inks must be avoided.
- The process is part of an automated line.

Laser marking has been or can be applied to semiconductor wafers, hand guns, gold, silver jewelry, precious stones and diamonds, chip capacitors, typewriter frames, railroad car wheels, consumer packages, heavy steel parts, large sheets of thin gauge steel, precision tool and die components and ceramic parts.

LASER PROCESSING OF SILICON

Laser are increasingly finding a home in silicon processing where electrons and radiations are replacing the wet chemistry and furnace treatment approach.

Lasers are applied to silicon for cleaning the surface, scribing the whole silicon wafers into individual chips, drilling holes for alignment, gathering (drawing impurities from the active front surface by damaging the back of the silicon wafer), annealing following ion implantation and to modify the crystalline structure and surface properties of silicon wafers.

FUTURE USES

The use of lasers in the industry is in its infancy. Special applications of small volumes are the general case. Lasers are generally accepted practices only in deep hole drilling and welding. As lasers of higher powers are becoming available and are installed on production lines, lasers will be more fully integrated and more diverse applications will arise.

Applications in automotive parts welding, cutting, drilling, surface treating, aircraft parts manufacturing and electrons industries are likely to grow the fastest in the near future.

Continuous CO_2 lasers represent the major use, they are commercially available in the 100 W to 20 KW and more range. The beam output is in the infrared region (10.6 nm) and is not visible.

The industry is likely to take advantage in many applications of some of the major advantages of laser:

- High power density, 10^6w/cm2 or more.
- Small size of the treatment area and low heat affected zone.
- Ease of positioning and accuracy.
- Non-attenuation of laser beam in air, hence the heat source can be removed from the workpiece.
- Simple protective shielding needed.
- Versatility and flexibility.
- Cost effectiveness.
- Reduction of scraps and rejects.

- Elimination of tool wear and replacement.
- Ability to increase throughput in a fully automated environment.
- Ability to drill very hard material.
- There is no contact between the "tool" and workpiece.
- Ability to treat or follow complex patterns and shapes.
- Edge quality cutting.

5 DIELECTRIC HEATING
Microwave, Radio Frequency Processes

INTRODUCTION

Dielectric heating includes *microwave* and *radio frequency* (RF) heating and is a process in which non-metallic materials are heated by absorbing high-frequency electromagnetic radiation. Heat is generated from within the object being heated. This method of heating contrasts the more conventional method of using convection and/or radiation to heat the surface of the object and then relying on conduction to transfer the heat into the object.

RF heating uses lower frequency radiation (i.e., wavelengths) than microwave heating, which allows it to penetrate deeper into the heated object and is generally used for thicker materials. Because microwave heating uses higher frequencies (shorter wavelengths) than RF heating, its heating intensity is greater and heating rates are faster. Dielectric heating is appropriate for heating electrically non-conducting materials which contain polar molecules, such as water. Thus, many dielectric applications are used to heat or dry moist materials.

RF and microwave heating are currently used in industrial applications. RF heating, being mature technology, has been used commercially since the 1930s. It is used for drying in the textile, food, lumber, paper and metals fabrication industries as well as for preheating and welding plastics. Microwave heating is more recent.

Advantages of dielectric heating include:

- Rapid heating because heat is generated internally.
- Control of heating rate (e.g., no warm-up time).
- Workpiece heated without heating the surrounds.

- Minimizes product deterioration due to significantly lower residence times.
- Less floor area than convection equipment.

Heat may be generated in electrically non-conducting materials by the absorption and dissipation of high-frequency electromagnetic radiation, commonly called dielectric heating. This radiation is in the approximate frequency range of about 300 to 300,000 MHz. In an effort to avoid conflict with communication applications using microwave frequencies, the Federal Communication Commission (FCC) has set aside several frequency bands for microwave heating. Allowed frequencies, 915 and 2450 MHz are used almost exclusively. Similarly, the most common used RF frequencies are 13.56 and 27.12 MHz. While in principle the heating mechanisms for both of these radiation spectra are the same, they have differences.

Dielectric heating can be used for any material comprised of polar molecules, i.e., molecules having an asymmetric electronic structure such that they tend to align themselves with an imposed electric field (water is an example). When the direction of the applied field changes rapidly, the molecules dissipate electric energy by molecular vibration in trying to keep pace with the alternating field direction. If the agitation is high, due to having a strong electric field (high voltage), the heating will be stronger. If the reversal of field takes place millions of times a second, the agitation is more frequent and the heat will also increase. All the molecules within the material that are exposed to the field will be agitated simultaneously. Thus, heat is produced throughout the material, rather than being imposed from the surface as in conventional heating. This difference offers significant advantages in some applications.

SYSTEM

All dielectric heating systems consist of the following components:

Generator: A power supply, voltage controls and a radiation source such as an oscillator (for RF) or a magnetron (for microwave). The power supply and voltage controls provide high voltage power to the radiation source, which generates high frequency power for the application.

Applicator: Transfers the high frequency power to the workpiece. RF heating, it houses the electrode system (which converts the high frequency power to RF waves), provides shielding and may include auxiliaries such as moisture extraction systems. For microwave heating, it consists of one or more waveguides to direct the microwaves from the magnetron to the product, and can also include one-way shields to prevent microwaves from reflecting back through the waveguide, possibly damaging the magnetron.

Materials Handling Equipment: Positions product under the applicator. In continuous processing systems, such as conveyors, the material is guided through the exposure area. Batch processing systems have no material handling system, so an operator must move the product.

System Controls: Includes the necessary controls (automatic, digital or manual) to regulate the processing exposure time, intensity and/or material handling speed.

Several different types of radiation sources are used. In the radio frequency range, simple triode oscillators are used. These low-cost devices have been built in unit sizes up to about 1.5 MW. Microwave radiation sources are more complex. Air-cooled magnetrons, available in ratings up to about 25 kW, are most commonly used for both industrial and residential/commercial applications. Magnetrons are basically a tube comprising of a rod-shaped cathode within a cylindrical anode. When power is supplied to the magnetron, electrons flow from the cathode to the anode, setting up an electromagnetic field. The frequency of the field is determined by the dimensions of the cavities which line the walls of the anode. When power is supplied to the magnetron, oscillators in the cavities form microwaves. A less common microwave radiation source is the water-cooled klystron which is available in ratings up to about 1 kW, used only to limited extent in very large industrial installations.

Applicators will vary depending on the application. Applicators for the RF range are generally less complex than those for microwaves, often consisting of simple parallel plate or parallel rod arrangements surrounding the material to be heated. For microwave applications, the most common type of applicator is the multimode cavity. This type consists basically of a metallic box of dimensions such that a number of resonant modes are produced at multiples of the imposed radiation frequency. This resonance produces a fairly uniform radiation field within the enclosure and is particularly well adapted to heating bulk products. For other geometries, such as thin sheets of filaments, special slotted wave-guide configurations have been designed to efficiently couple the microwave source to the target material.

Viewed as an electrical circuit, the applicator and the material being heated represent a capacitive load to the source. For efficient transfer of energy, it is critically important that the impedance characteristics of source and load be properly matched. This design problem is challenging, especially in microwave systems, since the electrical characteristics of many materials change during the heating process. In drying, for example, the dielectric constant of the material changes as water is driven out.

TECHNIQUES AND APPLICATIONS

The essential problem in the applications of radio frequencies is the transfer of energy from the generator to the product placed in an industrial environment. The efficiency of the generator being 60 percent and taking into account the high investment cost per usable kilowatt, the major part of the emitted energy must be absorbed by the product with acceptable uniformity. If this condition is not realized, we risk rapid deterioration of the equipment.

Certain applications are well known and the equipment is readily available, e.g., preheating of rubber, presses using HF for PVC and for wood, cooking, reheating, drying of textile and baking.

On the other hand, for new applications, the current approach for the design and installation of radio frequency equipment consists, in most cases, in the development of equipment specific for each case satisfying the global needs of the product and of the application.

Each new equipment is considered as a prototype requiring extensive testing which is translated in prohibitive investment costs and industrial risks, both for the manufacturer and for the user. We must compare this situation to the characteristics of the market for new applications.

These technologies are applicable to a large variety of products (food, paper, wood, plastics construction material, chemicals and pharmaceuticals) and processes (drying, heating, melting, polymerization and sterilization). However, in each industrial sector, because of high investment costs and very specific advantages, the use of radio frequencies is limited to a small portion of the possible products and processes. Many feasibility studies and tests have permitted the clear definition of the domains where the economic and technical balance sheet is positive.

This situation leads to various potential applications of comparable importance. Thus, the design of a piece of equipment applicable to only one application is an effort difficult to compensate, given the corresponding sales, especially when the equipment may not be provided by the same manufacturer. Consequently, we observe the specialization of manufacturers in one or two processes or products (drying of textile and solidification of PVC, rubber) which reduces the field of application of radio frequency techniques.

In all of the HF installations, there exists a matching network to permit the transfer of energy from the generator to the applicator containing the product; most often integrated into one or the other of these elements. Generally speaking, it is a case of insuring the best match of a load to a generator. Three functional components make up an HF installation; the generator, the matching network and the applicator; all intimately interlinked.

Generators are considered as elements of the circuit, and it is not necessary to study their construction. It is only necessary to know their output impedance and the nominal impedance of the load into which the generator delivers rated power at the working frequency. The output of the generator is a standard EIA coaxial connector. We can use the three types of generators: amplifiers, the working frequency is determined by a local quartz-controlled oscillator and the output impedance is independent of the load. This is not the case for the oscillators, where, in a certain sense, the output impedance depends on the power supply, on the load as well as on the operating point which make their analysis more difficult.

The calculation can be accomplished starting with the characteristics of the tank circuit at the nominal value of the frequency, current and voltage. We can then measure the output impedance with the help of a water load and calorimeter, a four-pole matching network and a frequency meter. This permits variation of the load impedances while measuring the dissipated power and frequency. When the nominal power

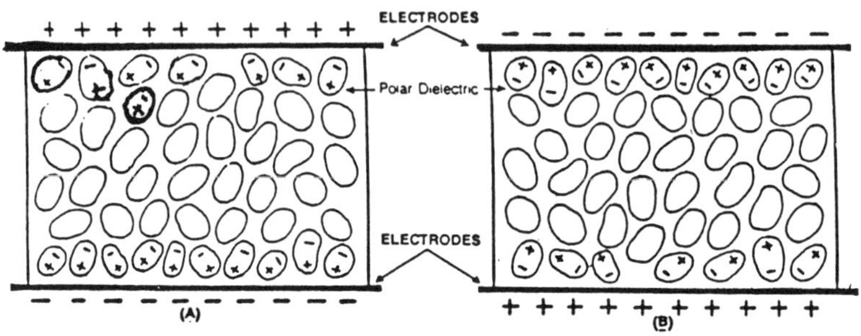

Figure 1. Polar molecules in an electric field rotate in synchronization with the field.

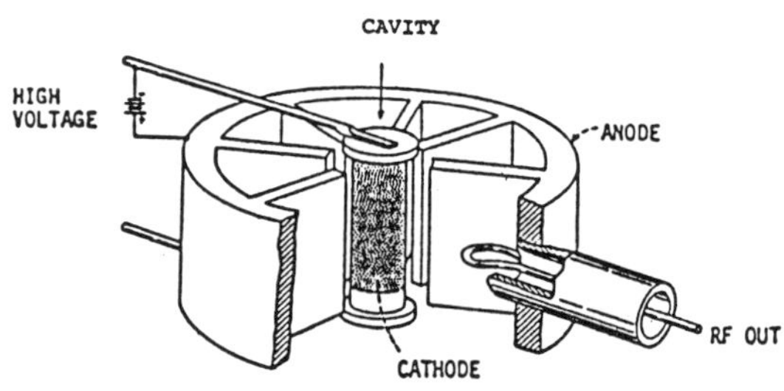

Figure 2. Magnetron.

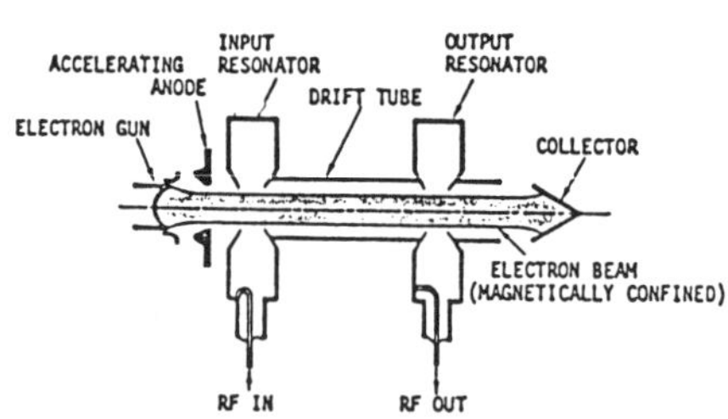

Figure 3. Klystron Magnifier.

is obtained at the nornimal frequency, the load impedance is disconnected from the generator, and its impedance is measured with a network analyzer.

Applicator is the part of the installation which is least reproducible from one application to another, since ideally it is determined by the product and the industrial application. In order to apply the concepts described to the design of HF installations, it is sufficient, to first order, to know the input impedance to the applicator. We have shown that this quantity is measurable and that it is possible to represent the applicator by an equivalent circuit facilitating the use of standard circuit analysis techniques.

Nonetheless, it is interesting to be able to calculate the electrical characteristic of the applicator by solving Maxwell's equations by numerical methods starting with the geometric configuration of the applicator and the material it contains, on one hand, to obtain the impedance; hence the possibility to predict the matching network and the information necessary for its construction beforehand and, on the other hand, to obtain the voltage and electrical field distribution in the applicator and to define the respective quality of the technical solutions with respect to the internal configuration. Such a calculating tool functions as an aid in the design of HF applicators allowing rapid evaluation of various possible solutions.

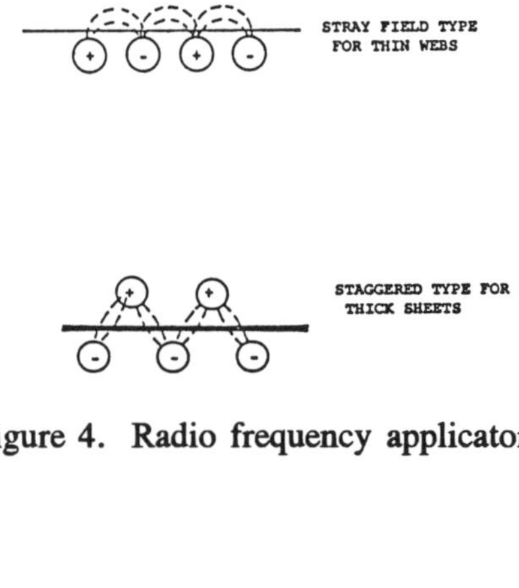

Figure 4. Radio frequency applicators.

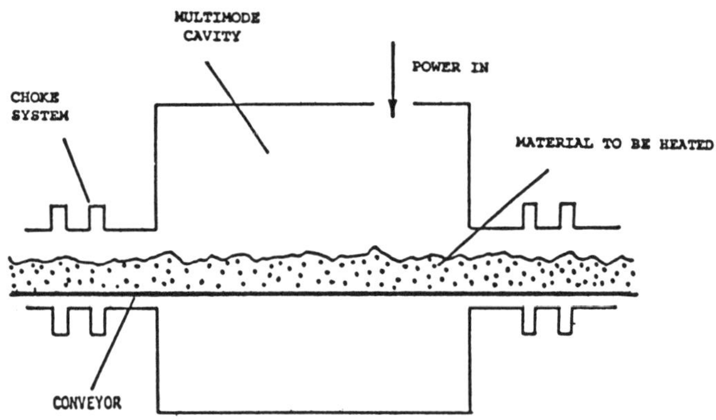

Figure 5. Multimode cavity microwave applicator.

Dielectric heating has much shorter drying times than convection drying due to the ability of microwaves and radio waves to heat objects volumetrically versus convection heating's dependence on conduction from the surface into the workpiece.

Microwave heating, demonstrates constant-rate drying. Microwave energy produces uniform vaporization throughout the material through the direct coupling of RF energy to the water molecules, resulting in rapid mass transfer from the material. If the proper frequency is chosen for the dimensions of the material being heated there is, strictly speaking, no limitation on the rate of heat and mass transfer achievable.
Heating occurs uniformly and mass transfer is limited primarily by the average power input to the material and the rate of convective removal at the surface.

The substantial reduction in drying time and the uniform heating characteristic of dielectric heating are the two most important advantages of RF energy. Speed and uniformity can result in accelerated production rates, reduced scrappage, higher product quality, lower energy and operating costs, reduced labor requirements, reduced floor space requirements, and a more desirable working environment than attainable with traditional combustion-heated equipment.

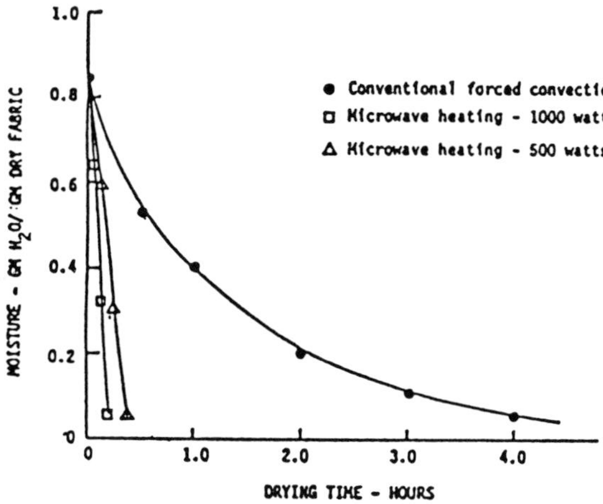

Figure 6. Moisture variation of fabric dried with microwave and conventional heating.

RADIO-FREQUENCY ENERGY

Radio-frequency energy is a source of heat which can be used as thermal energy in a variety of food processing operations. Frequencies above 300 MHz are the microwave region and their use is known as microwave heating. Frequencies below 300 MHz are known as the macrowave region and their use is known as dielectric heating.

The major molecules for radio-frequency heating is water. The greater the concentration of water in the material, the larger the dielectric loss factor for the material and the faster the product will heat up.

Microwaves are attractive for use in the air dehydration of some foods because of the speed of removal of moisture and the minimization of case-hardening. Applications enjoying reasonable use include the finish-drying of pasta and potato chips and the finish-baking of biscuits. The potential use of microwaves in freeze-drying has been described below. Experiments have proven the feasibility of using microwaves in foam-mat dehydration. A system has been developed by Pernod, in France, using microwaves in a vacuum tunnel to dry a great variety of solid and liquid food items. This system is more expensive than spray-drying but less expensive than freeze-drying. In a related application, the United States Energy Research and Development Administration has funded the development and testing of a microwave-vacuum system for drying grain.

A process using dielectric heating to concentrate liquids such as orange juice and apple juice has been used commercially since the 1960s. The Sargeant electronic concentration process uses radio frequency at 10 to 30 megacycles per second to evaporate water from a juice under high vacuum.

Radio-frequency heating in food concentration has enjoyed limited commercial use since radio-frequency energy is much more expensive than thermal energy from steam. The key to successful large-scale commercialization of radio-frequency energy processes is likely to be the

TABLE 1

COMPARISON OF ENERGY REQUIREMENTS FOR CONVENTIONAL AND MICROWAVE DRYING OF MACARONI

	Conventional	Microwave
Electricity usage (kWh/10^3 lb)	28 - 30	20 - 22
Electricity usage[a] thermal equivalent (10^6 Btu/10^3 lb)	0.30-0.32	0.21-1.23
Direct thermal (10^6 Btu/10^3 lb)	0.43-0.45	0.30-0.34
Total primary energy (10^6 Btu/10^3 lb)	0.72-0.77	0.51-0.57
Electricity cost (\$/10^3 lb at \$0.06/kWh)	1.70-1.80	1.20-1.30
Direct fuel cost (\$/10^3 lb at \$4.00/10^6 Btu)	1.70-1.80	1.20-1.36
Total energy cost (\$/10^3 lb)	3.40-3.60	2.40-2.66

[a]Based on conversion rate of 10,500 Btu/kWh

Source: Schmidt, Electricity and Industrial Productivity: A Technical and Economic Perspective, 1984.

production of a superior quality product, or the production of a unique product for which there is great demand.

Although the different electroconcentration processes examined are diverse, some general conclusions are apparent. In general, electroconcentration methods tend to be more capital-intense than conventional processes. Operating costs are somewhat more competitive with size of application a crucial factor. The energy-consumption competitiveness varies for different processes; some use much less energy than evaporation (e.g., reverse osmosis), while others use much more (e.g., microwave). The ability of reverse osmosis and ultrafiltration to decrease the BOD of plant effluents may greatly increase the market potential of those processes. Generally, however, in order for an electroconcentration process to substantially penetrate the food processing industry, it must provide a unique product in heavy demand or a product far superior to those of conventional concentration techniques.

WATER REMOVAL

Water is removed from foods for several purposes: to provide microbiological stability, to reduce chemical reactions which cause deterioration of food quality, and to reduce bulk for easier and more economical storage and handling.

A distinction can be made between dehydration and concentration processes based on the water content of the final product. Dehydration or drying processes produce a product with water content of less than 10% by weight; concentration processes produce a product with a water content no lower than 30% by weight. Dehydration is generally achieved by unsteady-state molecular diffusion from particles or droplets. Concentration is generally achieved through steady-state molecular and eddy transport from a fluid batch.

Major types of concentration processes include crystallization, clathration (partial crystallization followed by separation of crystals from concentrate), evaporation, osmosis, reverse osmosis, ultrafiltration, electrophoresis, and pre-evaporation (an evaporative process in which the liquid and gas phases are separated by a membrane that is selectively permeable to water). Food concentration processes must be inert with respect to the food products and must be selective, retaining all

TABLE 2

COMPARISON OF CONCENTRATION AND DEHYDRATION PROCESSES

Dehydration	Concentration
• Unsteady-state Process	• Stead-State Process
• Water Loss Via Molecular Diffusion	• Water Loss Via Molecular and Eddy Transport
• Performed on Pieces or Droplets	• Performed on Fluid Bath
• Product Water Content < 10 WT%	• Product Water Content >30 WT%

components except water. For foods containing aromas this selectivity is of special importance.

The conventional method for food concentration is evaporation using steam generated by the combustion of fossil fuels.

Processes which depend on electric energy for their driving force and which have demonstrated applications in concentrating foods are discussed. Those processes include ultrafiltration, reverse osmosis, freeze-drying, freeze concentration, microwave-drying and concentration by dielectric heating. For each process the principles and operating specifications are described in this book.

CONCENTRATION AND DEHYDRATION USING RADIO-FREQUENCY ENERGY

Radio-frequency (RF) energy is a source of heat which can be used as thermal energy for a variety of food processing operations including blanching, drying, pre-cooking, concentrating, pasteurizing and sterilizing, defrosting and cooking.

Macrowaves and Microwaves

The permitted frequencies in the U.S. for industrial, scientific and medical purposes are listed. These are the same frequencies as those permitted in Europe except for the 915 MHz frequency which is 896 MHz in Europe. The frequencies below 300 MHz are known as dielectric heating. The frequencies above 300 MHz are the microwave region and their use is known as microwave heating.

The power absorbed by a given material when placed into a radio-frequency field is a direct function of the dielectric properties of the material, the frequency of the field and the voltage gradient. The greater the frequency, the less the penetration of RF energy into matter. The choice of frequency for a particular application will depend on the cost of the power source and applying equipment for the particular frequency, the desired depth of penetration and the relative power input at the particular frequency.

Water is the major molecule for radio-frequency heating. The greater the concentration of water in the material, the larger the dielectric loss factor for the material, and the faster the product will heat up.

Microwaves in Dehydration and Concentration

Microwaves are used in the air dehydration of foods. A characteristic of radio-frequency energy is that when a food product is placed in a radio-frequency energy field, the energy "seeks out" the wettest material. The advantages are the speedy removal of moisture, particularly in the late stages of dehydration, and minimal case-hardening. Microwaves are being used for the finish-drying of pasta included in instant soup mixes. They are also used for finish-baking biscuits and finish-drying potato chips, both essentially dehydration processes.

Microwaves can potentially be used in freeze dehydration. The status of work in this area is described in a later section. Experiments have also been performed using microwaves in foam-mat dehydration. Foam-mat drying is a process used for dehydrating heat-sensitive products. In foam-mat drying, a product in liquid or semi-liquid form is mechanically whipped in order to inject air or other appropriate inert gas. The foam is then spread in thin layers on trays or conveyor belts where it is dried, ordinarily by exposure to drying air. Due to the increased product surface area, accelerated drying results. However, a limiting factor is the poor thermal conductivity of the drying foam. The use of microwave energy to heat the product increases the rate of drying up to ten-fold and allows thicker foam layers to be used. Foam-mat microwave drying tests for tomato paste, orange concentrate and onion

TABLE 3

FREQUENCIES PERMITTED FOR INDUSTRIAL, SCIENTIFIC AND MEDICAL PURPOSES IN THE UNITED STATES (AND ASSOCIATED WAVELENGTHS)

Frequency MHz	Wavelength (M)
13.56	22.2
27.12	11.0
40.68	7.35
915	0.328
2,450	0.1224
5,800	0.0517
22,125	0.0136

puree have been performed without noticeable deterioration in product quality. Microwave drying seems most useful in the initial warm-up and first dehydration periods. The use of microwaves in the foam-mat process is still in development stages.

A system has been developed by Pernod, the French manufacturer of aperitifs, which uses microwaves in a vacuum tunnel to dehydrate or concentrate heavy food slurries or whole food items. A conventional microwave processing tunnel made by Les Micro-Ondes Industrielles (LMI) has been adapted to hold a vacuum from 1 to 20 Torr. Air-lock systems have been added for loading and unloading food. The food material is dropped through the air lock onto a continuous stainless steel belt. As the material passes through the microwave field, internal water molecules instantly heat up to around 30°C. At this temperature and under vacuum the water foams out of the food rapidly. Flavors and aromas are not changed, however. The foam gradually becomes dry and after being broken up by a turning scraper, the dry product drops through a plastic chopper into a collector just below the vacuum drum. A nitrogen blanket can be introduced over the product if this is required.

The complete process takes 20 to 40 minutes, depending upon the food. The experimental unit at Pernod has a drying capacity of four to seven liters of water per hour with a final product moisture level of less than one percent. Incoming material can have a solids content as high as 60%. This can greatly reduce energy costs for drying and can allow glucidic additives to be included in fruit concentrates to enhance flavor retention. The dried product which results tends to have excellent rehydration capabilities even in cold water. The system permits the processing of extracts concentrated up to 83% dry matter. Only viscosity is a limiting factor.

Several problems encountered in the development process have been solved. Ionization of microwaves under vacuum was remedied by stepping up the charging frequency from 915 to 2450 MHz. Sharp projections in the tunnel, which caused higher field values, were eliminated. To further avoid high-electric field values the radiation passes through a resonance chamber at atmospheric pressure prior to entering the chamber.

Products which have been tested include asparagus, strawberries, cream, coffee, herb tea, mushrooms, milk protein, beet root, other fruits, onions, paprika, carrots, vanilla, licorice, various infusions, essential oils, beer, natural colorings, sucrose, dextrins, eggs and many pharmaceutical products. The system is best suited for processing cut, diced and whole products such as vegetables and fruits.

Economically, the system is rated between freeze-drying and spray drying. It is most useful and economical for materials which are not suited for conventional dehydration methods, materials with a very high dry-matter concentration and materials of fixed composition. Under all these conditions the cost for evaporating one kilogram of powder competes with other processes. Potential applications of the Pernod process could result from its association with preconcentration by vacuum evaporation or ultrafiltration or freeze concentration.

In a related application, the U.S. Energy Research and Development Administration has funded the development and testing of a microwave-vacuum system for drying grain. Early tests on the microwave vacuum-drying of corn indicated 38% less energy is consumed, at a temperature 78°F lower than conventional dryers, and 73% faster.

Applications of Macrowaves in the Concentration and Dehydration of Food

Macrowaves have been commercially used for concentrating liquids such as orange juice and apple juice and have been used experimentally to concentrate a number of other fruit juices, fruit purees and tea. The process was developed during the 1950s and in 1960 was used commercially for orange juice concentration by Ralph Sargeant, whose name the process bears.

The Sargeant electronic concentration process used radio-frequency energy at 10 to 30 megacycles per second to evaporate water from a product (juice) under a high vacuum. Water evaporates at a juice temperature around 75°F or less.

The Lakeland, Florida, plant of Universal Food Products, has commercially utilized the Sargeant process for producing a sevenfold orange juice concentrate. The juice is concentrated in a two-stage process. In the first stage it is concentrated to 55° Brix by conventional low-temperature steam evaporation. The juice is further concentrated in the second stage to 72° Brix or higher using the Sargeant unit. The commercial adaptation of the electronic process is illustrated in Figure 7.

In the second stage the juice is pumped at 7 to 10 psi through a swept-surface heat exchanger to the electrode, where it is flashed into the evaporator. Eight percent of the heat input is made at the swept-surface heat exchanger and twenty percent at the electrode. Fifteen passes are necessary to raise the product from 55° to 72° Brix. The conventional evaporation part of the process takes 45 to 60 minutes, while the electrode final concentration takes 10 to 12 minutes.

The commercial unit used in this process can produce 250 gallons of 75° Brix juice per hour (apple juice). If two additional swept-surface heat exchangers are added, the capacity can be doubled. The utility requirements of the unit include 2500 pounds of saturated steam at 110 psi per hour, 380 gallons per minute of water at 78°F and 200 kW of electric power.

General Economics of Radio-Frequency Processing

Several generalizations can be made on the economics of this process. Macrowaves are less expensive to produce and use than microwaves. The major cost factor in microwave processing units is the applicator, although the tube costs are also important. As previously stated, microwave energy is much more expensive than direct electric heat or steam. In 1966 one Btu from electricity cost 3 to 5 times as much as one Btu from steam. The key to the success of these processes is likely to be the production of a better quality or unique product.

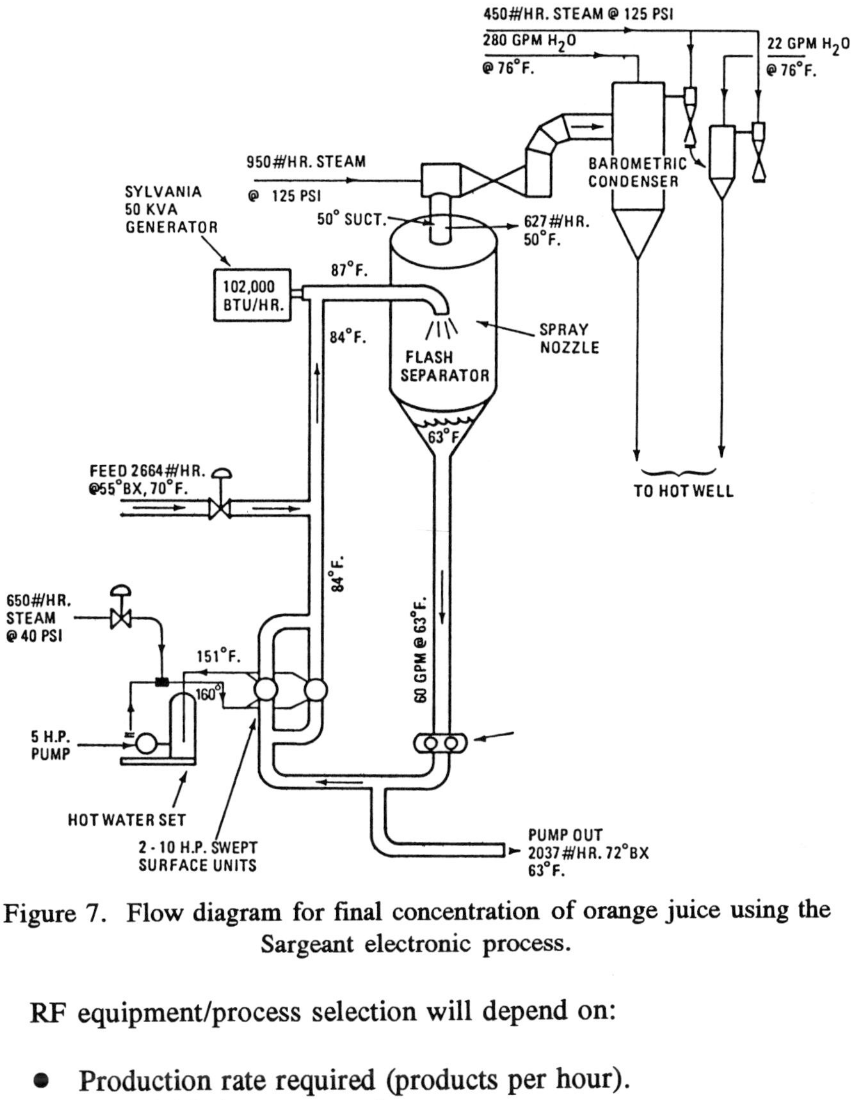

Figure 7. Flow diagram for final concentration of orange juice using the Sargeant electronic process.

RF equipment/process selection will depend on:

- Production rate required (products per hour).
- Material being processed.
- Weight and specific heat of product.
- Desired rise in temperature.
- Dielectric-properties (if application is not drying or heating water).
- Initial and desired moisture content (percentage of product weight).

One of the major disadvantages of dielectric heating is high capital costs. In general, RF systems cost less than microwave systems. System costs per kW vary based primarily on the type of operation. Continuous processing system requires a material handling system, as well as much more complex controls. Capital costs of an RF heater or dryer may range from $1000 to $3500 per kW. Smaller systems (1-200 kW) range from $2500 to $3500 per kW. Larger systems (300-1000 kW) range from $1000 to $2500 per kW. The high end of the ranges represent systems with sophisticated process controls and applicators, while the low end would be the cost of a simple applicator.

Capital costs of a microwave system ranges from $2000 to $4000 per kW. Again, the cost variation depends on the complexity of the system and whether the process is batch-type or continuous. For example, for a 40 kW unit, a continuous system would cost between $100,000 and $160,000 ($2500 - $4000 per kW), whereas a batch-type unit of the same size may cost only $50,000. Operating costs vary significantly depending on the specific application and plant situation.

Microwave Heating in Freeze-Drying

As a result of limitations on heat-transfer rates attainable in conventionally conducted freeze-drying, there has been some investigation of the provision of internal heat by microwave power. A schematic of one experimental unit used to investigate freeze-drying with microwaves is shown in Figure 8. Frequencies permitted for industrial applications in the United States are 915 MHz and 2450 MHz. In theory, microwaves should provide a very accelerated drying rate. This results from the fact that the heat transfer does not require internal temperature gradients that the ice temperature should be able to be maintained close to the maximum permissible frozen-layer temperature without excessive surface temperatures. For drying a 1-inch slab of frozen meat, a time of 2 hours of microwave drying compares favorably with about 15 hours for the conventionally dried slabs.

In spite of the apparent advantages, microwaves have not been successfully applied in freeze-drying. Major physical problems have not been resolved. One problem is a tendency toward glow discharge. This discharge, also called the corona effect, results in ionization of the gases in the chamber and undesirable changes in food quality. Useful power

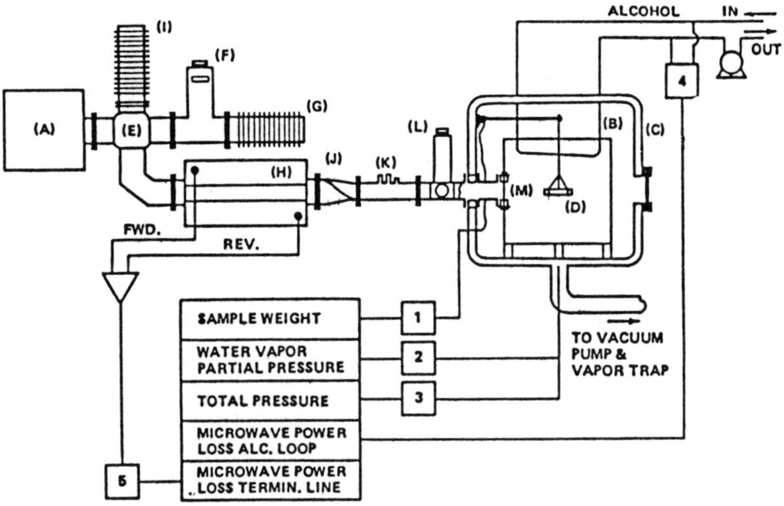

(A) Microwave generator
(B) Microwave applicator cavity
(C) Vacuum chamber
(D) Sample to be dried
(E) Four-port circulator
(F) H-tuner
(G) Dry resistor
(H) Bidirectional coupler
(I) Dry resistor
(J) Twist (Rotates polarization of electric field)
(K) 3-SUT
(K) 3-stub tuner
(L) E-H Tuner
(M) Wave guide vacuum transition
1– DC power supply, balancing unit, amplifier filler, stripchart, recorder
2– Hygrometer
3– Pressure gauge
4– Thermocouple temperature indicator
5– Power meter with thermistor mount

Figure 8. Schematic of experimental freeze drying with microwaves.

is lost as well. The glow discharge is greatest at pressures between 0.1-5.0 Torr and can be minimized at pressures below 50μ. Condenser operation at these pressures is quite expensive and produces a much slower drying rate. An additional physical problem is that the dielectric difference between water and ice can result in localized melting producing overheating. An additional factor which has impeded the development of microwaves for this use is that microwave energy is very expensive. It has been estimated that to supply one Btu for microwaves may require 10 to 20 times more energy than supplying one Btu from steam. Finally, microwave equipment suitable for large-scale continuous freeze-drying is not yet available on an economical basis.

TABLE 4

INTEREST IN MICROWAVE APPLICATION FOR INDUSTRY

Company	Applications of Microwaves
Air Products & Chemicals	Removing vinyl chloride from PVC resins
American Can Co.	Cross-linking polymers
ARCO Polymers Inc. (subs. Atlantic Richfield)	Curing flame-proofing compounds
Bendix Corporation	Bonding silicone elastomers
Branson Sonic Power Co.	Welding plastic
Celanese Corp.	Preheating UV-curable compounds
Chesebrough-Ponds Inc.	Sanitization of cosmetic color additives
CIBA-Geigy Corp.	Cross-linking polymers
Delavan Corp.	Sensor for liquid/solid level
Dow Corning Corp.	Curing silicone elastomers and foam mats
DuPont & Co.	Temperature measurement of nylon monofilament
Firestone Tire & Rubber	Vulcanization of urethane rubber
Ford Motor Co.	Sand and sodium silicate mold hardening
Goodyear, W.R. & Co.	De-vulcanization of sulfur-vulcanized elastomers
International Business Machines	RF sputtering apparatus
Lockheed Research Lab	Detoxification of phenyls, navy red dye and pesticides by plasmas
McDonnell-Douglas Aircraft	Soybean drying
MacMillan Bloedel Ltd.	Wood drying resin treatment
Microdry Corp.	Vegetable drying
Phillips Petroleum Co.	Drying carbon black pellets
Raytheon Manufacturing Co.	Tempering frozen meat
Scott Paper Co.	Graft copolymerization
Solar Turbines, International	Heating non-flammable polyannide foam

6 MATERIALS SEPARATION PROCESSES

The processes known as electrodialysis (ED), reverse osmosis (RO), ultrafiltration (UF) and ultracentrifugation (UC) may be characterized in general as material separation processes. Through these processes, dissolved substances and/or finely dispersed particles can be separated from liquids. The first three, electrodialysis, reverse osmosis and ultrafiltration, rely upon membrane transport - the passage of solutes or solvents through thin, porous polymeric membranes. The fourth process, ultracentrifugation, depends upon high centrifugal forces for separating dual liquids. A comparison of the characteristics of the four classes of separation processes is given in Table 1.

ELECTRODIALYSIS (ED)

The essential principle of electrodialysis is that electrical potential gradients will make charged molecules diffuse in a given medium at rates far greater than attainable by chemical potentials between two liquids, as in conventional dialysis. When a d.c. electric current is transmitted through a saline solution, the cations migrate toward the negative terminal, or cathode, and the anions toward the positive terminal, the anode. By adjusting the potential between the terminals or plates, the electric current and, therefore, the flow of ions transported between the plates, can be varied.

Electrodialysis lends itself readily to the continuous-flow type of operations needed in many industries. Multimembrane stacks can be built by alternatively spacing anionic- and cationic-selective membranes.

Among the technical problems associated with the electrodialysis process, concentration-polarization is perhaps the most serious. Other problems in practical ED applications include membrane scaling by inorganics in the feed solution as well as membrane fouling by organics.

TABLE 1

CLASSES OF SEPARATION PROCESSES

Name	Feed	Separating Agent	Products	Separation Principle	Application Example
Electrodialysis (ED)	Liquid	Anionic and cationic membranes; electric field	Liquids	Tendency of anionic membranes to pass only anions, etc.	Desalination of brackish waters
Reverse Osmosis (RO)	Liquid solution	Pressure gradient (pumping power) + membrane	Two liquid solutions	Different combined solubilities and diffusivities of species in membrane	Sea water desalination
Ultrafiltration (UF)	Liquid solution containing large molecules or colloids	Pressure gradient (pumping power) + membrane	Two liquid phases	Different permeabilities through membrane (molecular size)	Wastewater treatment; protein concentration; artificial kidney
Ultracentrifugation (UC)	Liquid	Centrifugal force	Two liquids	Pressure diffusion	Separation of large polymeric molecules according to molecular weight

Efficient separation of suspended matter from the influent stream, e.g., by activated carbon absorption, can reduce or prevent such problems.

The principal applications of electrodialysis include:

- Recovery of materials from liquid effluents, i.e., processes related to conservation, cleanup, concentration and separation of desirable fractions from undesirable ones.
- Purification of water sources.
- Effluent water renovation for reuse or to meet point source disposal standards required to maintain suitable water quality in the receptor streams.

Treatment of brackish waters in the production of potable supplies is the largest present application of electrodialysis procedures.

A broad range of costs is associated with electrodialysis processes, dependent upon such factors as the total dissolved solids (TDS), in the feed, the level of removal of TDS (percent rejection), and the size of the plant.

In the brackish water treatment, operating costs for very large ED installations (on the order of millions of gallons a day) are between 40 and 50 cents per 1000 gallons for brackish feedwaters. This compares favorably with RO costs.

A rule of thumb for the energy requirements for demineralizing 1000 gallons of saltwater by ED in large capacity plants (4 mgd) is 5 to 7 kWh per 1000 ppm of dissolved solids removed. Since the efficiency of electrodialytic demineralization decreases rapidly with increasing feed concentrations, this process is best utilized for treatment of weakly saline (brackish) waters containing less than 5000 ppm of total dissolved solids. In fact, for waters at the low concentration end of the brackish scale, ED may be the most economical process of all.

The second largest use of electrodialysis in the U.S. is the dairy industry, namely, the desalting of cheese whey.

Operating and maintenance costs for whey demineralization vary from about \$0.50 to \$1.90 per pound of product solids, corresponding to 50% to 90% removals. Electrical requirements may vary from 5 to 14 kWh per pound of product solids.

An increasing application of ED is the sweetening of prepared citrus juices. In a typical pilot-plant operation, electrical requirements were 0.22 kWh per gallon with a production rate of 95 gallons per hour.

Many other less extensive but nevertheless important industrial uses of electrodialysis are currently in commercial operation in the U.S. These include tertiary or advanced treatment of municipal sewerage water and treatment of industrial wastewaters such as metal plating baths, metal finishing rinse waters, wood-pulp wash water and glass etching solutions.

Potential applications of ED are numerous. A particular advantage of the electrodialysis process is its ability to produce solutions of high concentration of soluble salts. A combination of electrodialysis with conventional evaporation, for example, may be substantially cheaper than evaporation alone for the production of dry salt from saline solutions. Other potential application areas include:

- Replacing chlorine in the bleaching of wood pulp.
- Remineralizing dextrose greens and fish solutes in the food industry.
- Removing salts from canc sugar or potassium citrate from wine.
- Reducing salt content in the drug industry.

The eventual prospects of ED, in light of competing technologies such as reverse osmosis and crystallization, are difficult to evaluate. Certainly, continuing investment in ED systems, at least at today's rates, is expected to continue for some years in the future, especially under environmental and economic stimuli. With a few major manufacturers of ED equipment, there is not and probably will not be an appreciable degree of domestic competition in ED system production. In fact, there is little likelihood of any new major manufacturer entering the ED production market.

In general the ED market will evolve slowly as both pressures for environmental improvements as well as water reuse and material reclamation trends develop more fully. As some material prices have increased significantly, for example, it has become good economics to recover certain substances from dilute effluent streams which were previously disregarded.

REVERSE OSMOSIS (RO)

RO is fundamentally a means for separating dissolved solids from water molecules in aqueous solutions through membranes composed of special polymers which allow water molecules to pass through while holding back most other types of molecules. In the RO process the feedstream is split into a purified portion called the product water or permeate, and a smaller portion called the concentrate, containing most of the impurities in the feedstream. The percentage of product water obtained from the feedstream is termed the "recovery." Some of the important advantages of the reverse osmosis process include:

- Low energy consumption is characteristic. Because no change of phase is involved (as in distillation), the principal source of energy consumed is electrical (or possibly steam) to drive pumps.
- Relatively simple processing equipment is required, resulting in low-to-moderate equipment costs.
- Operations at ambient temperatures minimize scale and corrosion problems.

Major problems inherent in general applications of RO systems include the presence of particulate and colloidal matter in feedwater, the precipitation of soluble salts and problems associated with the physical and chemical makeup of the feedwater.

Current applications of large-scale reverse osmosis systems include:

- Upgrading (desalination) of brackish feedwaters to produce quality water supplied for municipal and/or industrial requirements or resort communities.
- High-purity rinse water for the solid-state electronic components manufacturing industry.
- Production of boiler feedwater and process water.

The reverse osmosis process appears to have the economic lead in relation to energy consumption for the treatment of all but the most saline waters. Typical energy consumptions for treatment of brackish waters fall in the approximate range of 7 to 12 kWh per 1000 gallons. Comparable energy costs for the most efficient desalination plants employing distillation range between 1.5 to 2 times higher per 1000 gallons. RO costs increase with increasing feed concentration and operating pressure and decrease with increasing efficiency of recovery (rejection).

For municipal wastewater treatment applications, an activated sludge system coupled with an RO system can cost slightly more than a conventional treatment process for a 100 million gallon a day (mgd) capacity (41 vs 35¢ per 1000 gallons), about the same for 10 mgd, and somewhat lower for 1 mgd ($1.22 vs $1.55 per 1000 gallons). Treatment effectiveness is greater with the RO system.

Other smaller-capacity reverse osmosis operations include:

- Treatment of industrial wastes, including metal finishing effluents, plating rinses and refinery wastewater.
- Treatment of wastewater and the retrieval of protein substances from fish and shellfish processing plants.
- Processing of pulp and paper mill effluents for reuse.
- Waste treatments in food processing in the dairy products industry in which edible protein and lactose are recovered from cheese whey.
- Desalination (including boron removal) of irrigation return flows.
- Food concentration applications.

Membrane processes, particularly RO, show considerable promise in the food processing industry. Many liquid foods are rendered susceptible to damage either by evaporation itself or by the effects of high temperature. Reverse osmosis and other membrane processes provide a method of concentrating and processing foods economically, without detrimental heating or phase change.

For RO, the potable water production market is estimated at $27 million and the maximum potential market size may be as high as $900 million. Brackish aquifers underlie much of the U.S.; if RO proves to be an economical means of creating potable water from these sources, a considerable market would develop in currently water-short areas of the country.

High-purity rinse water in the semiconductor industry and boiler feed-water production represent markets. Both of these markets are estimated at 10% capture. Industrial and municipal wastewater reclamation represents a large potential growth area. For all applications of RO systems, actual market size is estimated at $44 million, and maximum market size over $1 billion. Other major potential markets are in pulp and paper, whey concentration, and plating and metal finishing.

Less significant and more distant markets are in acid mine drainage, petroleum refining and corn milling operations.

ULTRAFILTRATION (UF)

In an ultrafiltration process, a feed emulsion is introduced into and pumped through a membrane unit; water and some dissolved low-molecular weight materials pass through the membrane under an applied hydrostatic pressure. In contrast to ordinary filtration, there is no build-up of retained materials on the membrane filter.

A variety of synthetic polymers, including polycarbonate resins, substituted olefins and polyelectrolyte complexes, have been employed for ultrafiltration membranes. Many of these membranes can be handled dry, have superior organic solvent resistance and are less sensitive to temperature and pH than cellulose acetate, widely used in RO systems.

In UF, molecular weight (MW) cutoff is used as a measure of rejection. However, shape, size and flexibility are also important parameters. For a given molecular weight, more rigid molecules are better rejected than flexible ones. Ionic strength and pH often help determine the shape and rigidness of large molecules.

Operating temperatures for membranes can be correlated generally with molecular weight cutoff. For example, maximum operating temperatures for membranes with 5000-10,000 MW cutoffs are about 65°C. For a 50,000-80,000 MW cutoff, maximum operating temperatures are in the 50°C range.

The biggest industrial use of ultrafiltration is the recovery of paint from water-soluble coat bases (primers) applied by the wet electrodeposition processes (electrocoating) in auto and appliance factories. About 500 installations of this type are operating around the world. The recovery of proteins in cheese whey (a waste from cheese processing) for dairy applications is the second largest application, where a market for protein can be found (e.g., feeding cattle and farm animals). Energy consumption at an installation processing 500,000 pounds per day of whey would be 0.1 kWh per pound of product.

Another large-scale application is the concentration of waste oil emulsions from machine shops, which are produced in association with cooling, lubrication, machining, rolling heavy metal operations, etc.

Ultrafiltration is a preferred alternative to the conventional systems of chemical flocculation and coagulation followed by dissolved air flotation. Ultrafiltration provides lower capital equipment, installation and operating costs.

Ultrafiltration of corrosive fluids such as concentrated acids and ester solution is an important application. The chemical inertness and stability of ultrafilters make them particularly useful in the cleaning of these corrosive solutions. Current uses include separation of colloids and emulsions, and recovery of textile sizing chemicals.

Biologically active particles and fractions may also be filtered from fluids using ultrafilters. This process is used extensively by beer and wine manufacturers to provide cold stabilization and sterilization of their products. It is also used in water pollution analysis to concentrate organisms from water samples.

Food concentration applications can be applied to processing milk, egg white, animal blood, animal tissue, gelatin and glue, fish protein, vegetable extracts, juices and beverages, pectin solutions, sugar, starch, single-cell proteins and enzymes. These areas represent a large potential growth market for UF.

Commercially, ultrafiltration equipment has been on the market since 1968. In 1970, sales were approximately $2 million. In 1974, sales were about $9 million. The 1975 market size did not grow from 1974 levels due to a general contraction of the economy and industrial processing. Paint recovery currently accounts for about 20% of capital expenditures, and recover of proteins in cheese whey and concentration of oily emulsions represents 60% of the market. The other 20% of the UF business is in proprietary areas. Estimates are for a market of $14-15 million, or a growth of $1 million/year. An upper limit for this market has been estimated between $30 million and $40 million, but this potential will not be fully attainable unless we are forced to adopt ultrafiltration as a control technology.

ULTRACENTRIFUGATION (UC)

Ultracentrifuges utilize intense gravity forces generated by the centrifugal forces of rapid rotation. These high-g forces are used to determine sedimentation coefficients and to separate subcellar fractions such as protein which may be relatively close in mass.

TABLE 2

ENERGY CONSUMPTION AND CAPITAL COST COMPARISON FOR MATERIAL SEPARATION PROCESSES

Process	Energy Consumption	Capital Costs
Electrodialysis	5-7 kWh/1000 ppm dissolved solids removed per 1000 gal	\$500-1500/gpm; \$15-25/ft^2 of membrane
Reverse Osmosis	7-12 kWh/1000 gal brackish water	\$15-30/ft^2 of membrane
Ultrafiltration	3-6 kWh/1000 gal	\$300-600/gpm

The ultracentrifuge has been an invaluable research tool in the differential separation of particulate suspensions and in the study of sedimentation rates. It is utilized extensively in the fields of virology and cancer research to separate viral particles from solutions. The introduction of moving-gradient centrifugation permits large volumes of virus-containing fluids to be processed, resulting in all the virus particles sedimenting into a narrow band. This is particularly useful in the preparation of vaccines.

Zonal centrifuge rotors have a variety of shapes and descriptions and can be classified in several ways. The simplest distinction is to call those rotors which hold a fixed volume of sample as "batch-type" and those rotors in which the sample can be either a fixed volume or continuously varied as "continuous-sample-flow-with-isopycnic-banding" (flo-band) type. While most batch-type zonal rotors can be operated in existing preparative-type centrifuges, the newer flo-bank rotors have a different shape and configuration and require special centrifuge drive systems.

Actual uses of UC include purification and separation of viruses, testing of semiconductors and protein separation in the manufacture of cosmetics, as well as protein separation from human and animal serum. Potential uses include protein or viral and bacterial separation in special food applications, and the analysis of water samples for biomass, pesticide contamination, and minerals.

In the past, several problems have tended to limit the usefulness of the ultracentrifuge. Ultracentrifuge cells are subjected to large forces and have been prone to problems of leaking. Rotors and bearings deteriorate with use, requiring that the ultracentrifuge be used at less than its rated maximum rotational speed. Large stresses and forces limit the useful lifetime of the ultracentrifuge. Ongoing research programs dealing with high-strength materials, oil-powered drives and high-pressure seals are attempting to alleviate these problems. They will have to be solved before the ultracentrifuge will be extensively utilized by industry. Currently, the turbines have to be exchanged every 2500 operating hours, although the entire unit itself may last for many years.

The greatest factors holding back ultracentrifugation from industry in general are that it is applicable to relatively low flow rate processes, and it is energy and equipment intensive.

7 FREEZE CONCENTRATION

INTRODUCTION

Freeze concentration is a separation process for liquid mixtures (solvents). It separates liquid mixtures by removing heat from the mixture until one or more components solidifies or forms a crystalline phase and then physically separating the crystals from the mixture by mechanical methods. The separated crystals are usually very pure, except for minor contamination from the liquid layer adhering to the crystal surface. The crystals are usually melted with heat recovered from precooling the mixture. Freeze concentration is an alternative separation process to evaporation, distillation and membrane filtration. The advantages of freeze concentration over distillation and evaporation include:

- Lower energy consumption. The energy required to crystallize liquid components is much less than energy required to vaporize them.
- Lower temperature operation. Fewer corrosion and/or scaling problems, thus permitting the use of cheaper construction materials.
- Improved product quality. Since the fluids are not subjected to high temperatures, products do not experience heat damage.
- Full retention of volatile components. This is often critical to such product characteristics as taste and functionality.

PRINCIPLES

Three major steps in the freeze concentration process are:

- Freezing.
- Crystal Recovery.
- Melting.

A very simplified block diagram of an indirect freezing process, showing the general flow through the three major steps is illustrated in Figure 1. A more detailed process diagram, illustrated in Figure 2, shows the components and flows of a typical freeze concentration system. A general operating mode is as follows. Feed enters the process through heat exchangers where the cold is "recovered" from the product and concentrate streams exiting the system. This feed then enters the freezer or crystallizer where a portion of it is converted to a solid, in discrete, relatively small (100 to 500 microns) crystals. The crystals are pumped as a slurry from the freezer to a separator which removes the liquid from the slurry, and usually washes the crystals with a small portion of either melted product or some suitable wash material that will not dissolve the crystal. Most of the liquid is recycled to the freezer where more of the desired constituent is recovered. Excess concentrate is bled from the system through the feed exchanger, to prevent excessive buildup of impurities. The washed crystal is then melted, either directly in contact with the refrigerator (which condenses on the crystal, giving up its latent heat to melt the crystal), or through a heat exchange surface. Part of the melted product can be recycled to reslurry the washed crystal, as necessary. The net product flow is pumped through the feed exchanger and out of the system. The refrigeration equipment removes heat from the freezer and transfers it to the melter, with any excess heat being removed from the system through the heat rejection compressor.

For applications in which the crystals of different species have similar physical properties, the methods of segregating crystals may not be adequate. It may be necessary to repeat the process, recycling the mixture through several crystals. Staging is an effective way to improve yield and decrease energy use.

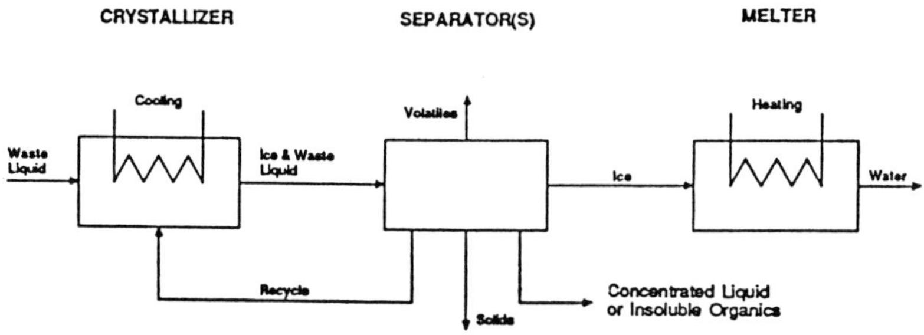

Source: Heist Engineering Corp., Industrial Applications of Freeze Concentration Technology, 1987.

Figure 1. Block diagram of indirect freezing process.

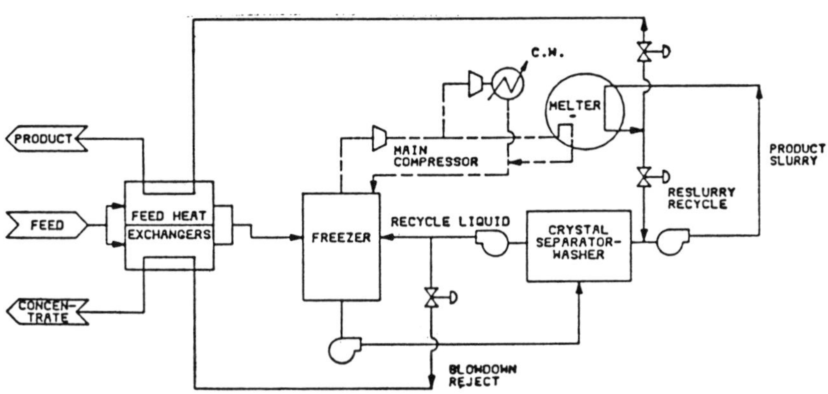

Source: Heist and Barron, "Freeze Crystallization Processes: Efficiency By Flexibility," 1983.

Figure 2. Process diagram of freeze crystallization.

Equipment and System Configuration

Freezing takes place in a vessel called a crystallizer. One type of crystallizer can be a scraped surface heat exchanger, which is a very expensive component. An alternative crystallizer is the seeded slurry falling film heat exchanger, which uses flow velocity to keep the heat exchanger tube(s) clear. Both crystallizers are indirect types, i.e., the refrigerant is kept separated from the liquid mixture by a heat exchange surface. In many applications the refrigerant can be mixed directly with the process solution, thus eliminating the need for a heat exchanger. The freezer is then a type of mixing chamber.

Crystal recovery involves separating the mixture (and refrigerant, if used) from the crystals and separating the crystals from one another if various types are formed. Although the crystals can be very pure, they usually have some minor contamination from the liquid layer adhering to the crystal surface. In some cases, a washing operation is needed to remove the liquid and wash the surface layer from the crystals. The most efficient and lowest cost washing method uses a wash column, which is essentially a packed bed through which the liquid flows, separating out the crystals. When more than one type of crystal forms, the individual crystals must first be separated from each other for recovery of pure components. This separation can be done to varying degrees in crystal segregation devices, operating on differences in physical properties or crystal sizes. Centrifuges and specially designed differential sedimentation tanks can be used for this purpose.

Melting of the crystals converts them into a useable form and also recovers the heat removed by the crystallizer. The melter-condenser can be either a shell and tube heat exchanger or a direct contact packed bed.

As previously mentioned, there are two basic types of freeze crystallization processes, indirect and direct, depending upon whether the refrigerant comes into contact with the mixture in the freezer vessel. Direct processes are further classified according to whether the refrigerant is one of the primary solution components, whether the refrigerant is a separate (secondary) material, and whether the crystals formed are of the pure substances or are clathrates (solid solutions having more than one material in the crystal).

Indirect systems have a heat exchange surface between the process mixture and the refrigerant. These systems have been used for over 40 years. The main differences between the various indirect freezers are the use of different refrigeration cycles and the different methods of keeping the heat exchange surface clear of ice deposits. Surface clearing can be accomplished by mechanically scraping the ice deposits from the heat exchanger as they form, by pumping scraper balls through the heat exchanger tubes, or by maintaining high velocity flows that prevent ice build-up in the first place.

Direct-Primary or Triple Point systems involve the simultaneous vaporization and freezing of the process mixture. Some of the fluid is vaporized, taking heat from the remaining mixture, thereby cooling it and causing ice to form. The mixture operates near its triple point. A large compressor or an absorbent loop removes vapors from the freezer vessel. Absorbing the vapor into a liquid or freezing the vapor onto a much colder surface are recently developed methods of removing the vapor, resulting in smaller volumes of materials needing to be transferred out of the freezer vessel.

Direct-Secondary systems add a separate refrigerant to the freezer vessel and then vaporize the refrigerant, thereby crystallizing part of the process mixture. Far less material transfer occurs in direct-secondary systems compared to direct-primary systems, but crystal growth vessels, wash columns, decanters and stripper columns are used to help separate, strip and recover the refrigerant. Some commonly used refrigerants include butane, propane and various fluorocarbons.

Direct-Clathrate systems also add a separate refrigerant, but use one that forms crystals having two or more pure compounds bound in the matrix (similar to metal alloys). The refrigerant and solvent molecules join physically, but not chemically, causing the solvent crystals to exist at temperatures higher than if the crystal were purely solvent. Consequently, direct-clathrate systems have the additional complexity over direct-secondary systems of having to separate and melt the clathrated materials. A main advantage of this system is that it operates at near-ambient conditions, thus making it feasible for many separations that other freeze processes cannot perform well.

The selection of a specific type of freeze crystallization cycle (i.e., indirect, direct-primary, direct-secondary or direct-clathrate) depends on the type of application, process requirements, technology availability and capital and operating costs. For example, indirect and direct-primary cycles are suited for applications where refrigerant cannot be mixed with process fluids because of product quality and/or health considerations.

The main advantages of freeze concentration over the conventional separation processes are reduced energy consumption, fewer corrosion and scaling problems, and improved product quality, including complete retention of volatiles and aromas.

The freezing process is less energy intensive than evaporation or distillation, but depending on the products involved, the system type and the equipment design, not all of the difference in energy required will be realized as energy savings. For example, the latent heat of fusion of ice (i.e., the amount of heat removed from water to change it to ice at the same temperature) is only about 144 Btu/lb, while the heat of vaporization of water (i.e., the amount of heat required to vaporize water) is 970 Btu/lb. But if an indirect freeze system with a scraped surface heat exchanger crystallizer is used, then the energy required to operate the refrigeration compressor, crystallizer and auxiliary equipment could exceed that required for evaporation. This is because scraped surface heat exchangers have low heat transfer coefficients, they must be sized with very large temperature differentials. But if a direct freezing system were used, the energy requirement for the refrigeration equipment could be 1/2 or 1/3 of that needed for the indirect system.

FREEZE DRYING

Freeze-drying as a process which in its most common usage is not really electro-concentration. Water is vaporized from frozen food under vacuum, without passing through the liquid phase. The heat for subliming the ice may be provided in a variety of ways but the most common is by radiation from shelves warmed by a heat-transfer fluid which has been heated by steam.

In theory, the heat for sublimation of the ice can be provided by microwave energy and, in fact, the use of microwaves should be expected to accelerate drying considerably. Substantial research has been performed on this application of microwave energy but major physical problems remain to be resolved. A major problem is the glow discharge or corona effect which results at pressures between 0.1 and 5.0 Torr, which results in ionization of gases in the chamber and undesirable changes in food quality.

The capital and operating costs of freeze-drying are comparatively high, thus limiting its adoption to those applications in which significant quality improvements are realized over conventional processing. Microwave energy is very expensive in its own right. As a result, in addition to the physical problems, microwave freeze-drying is probably quite far from commercialization.

The major commercial use of freeze-drying is in the production of instant coffee. Because the freeze-dried coffee is of significantly better quality than other instant forms, its use for this application has become rather widespread. Numerous other applications are in use, but none as extensively as freeze-drying coffee.

FREEZE CONCENTRATION

The freeze-drying of coffee is often preceded by a preliminary electro-concentration step known as freeze concentration. In this process the coffee is frozen into a slush, usually in scraped-surface heat exchangers, and the concentrate is removed from the ice crystals mechanically. A press, centrifuge or wash column may be used for the separation. The most common separation method in the use is the centrifuge.

Freeze concentration systems have relatively high capital costs. The freezing process requires fewer Btu than evaporation. However, when the energy consumption of the separation process is considered, only a minimal energy advantage results. A problem that limits commercial application is loss in yield due to entrapment of solute in the ice crystals. Because of this, the freeze concentration of coffee before freeze-drying is the only application currently realizing sizable usage. Limited usage is reported in the concentration of vinegar.

The Dutch firm, Grenco, developed a freeze concentration system which promises to solve some of these problems. The system has lower capital costs than others and controls crystallization so as to minimize the captured solute. This process has been used in Holland to concentrate coffee, apple juice, orange juice and other products. It is not yet on the market in the U.S., but its introduction is anticipated.

Economics

The cost of freeze concentration separation depends upon the type and quantity of produce involved, as well as the type of freeze system used. Cost estimates by Freeze Technologies Corporation (formerly Heist

Engineering) for a 'standard' direct-secondary system range from $0.10 to $0.30 per gallon of liquid waste treated for a 10 gpm system (this estimate includes both equipment amortization and operating expenses). Because of the significant dependency of cost estimation on the specific application, this economic section will only present ranges of costs and some specific examples of costs, though no 'typical' costs.

The capital cost for freeze concentration systems is high, ranging from 1.5 to 7 times as much as a conventional evaporation system. One of the main reasons for the high capital cost is that each concentration system is custom designed and built. The factors that affect the design and cost of a freeze plant are the following:

- Size or capacity required.
- Materials of construction.
- Freeze system type.
- Viscosity of the process stream.
- Freeze temperature required.

A 'standard' direct-secondary freeze system producing 10 gpm of clean product water from hazardous waste water and that comprises a transportable and highly automated unit would cost $1.0 to $1.5 million. Relative to the direct-primary system, the capital cost of the direct-clathrate system is 10% more, the direct-primary 20% more, and the indirect system from 50% to 150% more. A scale-up exponent of 0.75 for the indirect system and 0.6 for the direct systems can be used to estimate the capital costs for other sized systems, (i.e., the cost for a larger system can be estimated from the ratio of the plant sizes raised to the scale-up exponent, times the cost of the 10 gpm system).

Indirect systems inherently have higher capital costs due to the complexity of the equipment. Indirect freezing requires a heat exchanger and scraped surface types require apparatus to perform the scraping. Energy costs vary depending on the type of freeze system. Indirect freeze systems require the most energy due to lower efficiencies and additional equipment. A major operating cost is labor, which varies with the size of the system and the management of the facility. Labor costs for freeze systems are comparable to those for conventional systems. Power, supplies and refrigerant makeup, and maintenance costs are generally about three equal costs.

In terms of system sizes, indirect, direct-primary and direct-secondary systems have been designed to handle feedstock capacities of over 35,000 pounds per hour in either commercial units or pilot plants. Direct-clathrate units, though, have not yet operated above 200 pounds per hour. In addition, indirect and direct-secondary systems have been designed at capacities of 50,000 lbs/hr for para-xylene fractionation. But, in general, indirect systems are not particularly efficient, nor do they scale well to applications with very large heat loads. On the other hand, direct contact freezers have no practical or theoretical size limitations.

The most promising opportunities for freeze processing exist within four manufacturing industry groups, all of which process volumes of liquid mixtures:

- Food.
- Pulp and paper.
- Chemicals industry.
- Petroleum industry.

To date, freeze concentration has been most extensively applied in the food and beverage industry, which increasingly is taking advantage of the superior product quality achievable with this technology. Numerous plants are now in operation in the U.S. and abroad for concentrating orange, grapefruit, apple and other fruit juices. Other applications include the concentration of beer, wine, vinegar, sugar solutions and other food liquids. Application of the technology to dairy products is now under development, specifically milk dehydration and cheese whey concentration. Still another major use of freeze concentration is for the concentration of coffee prior to freeze drying or spray drying.

In the pulp and paper industry, much of the energy consumed is used to concentrate the liquid stream, known as "black liquor," resulting from the chemical separation of wood fibers in the pulping operations. In current practice, multi-effect evaporators are used to concentrate the black liquor, which is burned to recover inorganic chemicals for reuse and to utilize its heat value to generate electricity and steam. Freeze crystallization was first proposed in the 1970s as a means of preconcentrating black liquor prior to the multi-effect evaporators. It was shown that black liquor could be concentrated by indirect and direct-primary freeze crystallization. But, the use of freeze crystallization and other processing steps to replace multi-effect evaporation is still under research.

In the chemicals industry, freeze concentration is used in the production of para-xylene and is potentially applicable to a wide range of other chemical processes, such as concentration of acetic acid and of caustic (sodium hydroxide). The process has also been applied to desalination and has potential application to many other wastewater treatment tasks. Considerable research is proportional to the price of crude oil and residual fuel oil.

Installation/Applications

To date, the commercial successes of freeze concentration have primarily been with indirect systems. These successes have occurred where the technology is better able than the heat-driven alternatives to perform the desired separation (e.g., dewatering of coffee extract prior to freeze drying).

The following are recently installed or soon to be initiated installations of freeze concentration.

- Before the end of 1989, a 10 gpm waste reduction by freeze concentration demonstration plant was scheduled at the Stringfellow site in Riverside, CA. This site is ranked 7th on the national priority list of EPA Superfund sites. The system was designed by Freeze Technologies Corp. and built by Applied Engineering. The plant currently resides in Orangeburg, SC. A visitors day will be scheduled when the Stringfellow plant is operational. The system is a direct-secondary process.
- In March 1988, the C. W. Nofsinger Co. (Kansas City, MO)--the U.S. license of Tsukishima Kikai's (Tokyo) freeze concentration technology--made its first sale, a 36-million-lb/yr paradichloro-benzene plant, to Monsanto Chemical Co., in Sauget, IL. The unit was due to be onstream by mid-1989. The Tsukishima Kikai process has been used since 1983 in Japan.
- In April 1988, CBI Freeze Technologies Inc. (Plainfield, IL) shipped a truck-mounted 1500 gal/d freeze concentration plant. Bound for a Chemical Waste Management Inc.

facility in Kettleman Hills, CA, this is the first unit to use freeze concentration technology for treating hazardous wastewaters.

- Calyxes Research and Development Corp. (Albuquerque, NM) says that it will "soon deliver" the first direct-contact, primary-refrigerant freeze concentration system, for purifying 20,000 lb/d of organic chemicals at Unique Chemtech (Albuquerque). Later, two 250,000 gal/d units will be delivered for customers processing mining and oilwell waste streams.
- The Dutch firm Grenco Process Technology B.V. ('s-Hertogenbosch), has unveiled a six-stage system, the W-100. Grenco says the new design has half the capital costs of its previous model.
- A French firm, Societe Nouvelle des Etablissements Olier (Clermont Ferrand), plans to market a fruit-juice concentrator based on a new technique developed by Institut de Recherches sur Les Fruits et Agrumes (IRFA), part of the Institut National de la Recherche Agronomique (both in Paris). Olier is not willing to divulge more details, pending resolution of some patent issues.
- Finally, Switzerland's Sulzer Bros.' (Winterhur) freeze concentration system for purifying organic chemicals got its first commercial tryout in North America. A sales manager at Sulzer Canada Inc. (Rexdale, Ont.), said that the firm had sold freeze concentration plants with a large combined capacity. They expect sales to continue on its upward trend.

Freeze concentration systems have been commercially used in a few applications, predominantly indirect systems, but have proceeded through pilot testing in numerous other applications. A summary of the commercial and pilot plant freeze systems developed is shown in Table 1.

Current research and development of freeze systems is being conducted predominantly by freeze equipment manufacturers, and mainly focusing on developing specific process systems for specific applications, though development in freeze system-specific equipment, such as wash

TABLE 1

FREEZE SYSTEM DEVELOPMENT STATUS

Freeze Cycles	Status P	Status C	Application	Size (#/Hr.)
Indirect		X	Fruit juice, beer, coffee acetic acid concentration	10,000
		X	P-xylene and other organic materials fractionation	50,000
		X	Seawater desalting (in start-up phase)	35,000
Direct Primary	X		Seawater desalting using vapor compression cycle (1-yr successful test)	35,000
	X		Seawater desalting using absorbtion cycle (1-yr successful test)	7,500
	X		BTX and other organic materials fractionation (lab scale tests and sample recovery)	100
Direct Secondary	X		Seawater desalting (several successful tests)	35,000
	X		Wastewater treatment (lab & field-scale successful tests)	100
		X	P-xylene fractionation	50,000
Direct Clathrate	X		Seawater desalting (several test series, with success in demonstrating concept)	100
	X		Crystallization & eutectic separation of sugar, black liquor and other materials	200

Notes: P = Pilot plants
C = Commercial units

Source: Heist Engineering Corp., Industrial Applications of Freeze Concentration Technology, 1987.

columns and decanters, and in simulating components and the entire freeze system is also being conducted. The leader in the direct freeze concentration development is Freeze Technologies Corporation (a joint venture of Heist Engineering Corp. and Environmental Systems Corp.), and their primary focus is in waste reduction applications.

There are a handful of firms involved in the designing of freeze concentration systems and the manufacturing of freeze concentration equipment. Designers of indirect freeze systems or manufacturers of components include Anco/Votator, CB&I (Chicago Bridge & Iron) Freeze Technologies, Inc., Grenco Process Technology B.V. (Holland), C. W. Nofsinger Co. (Kansas City, MO), Phillips, Struthers, Sulzer Canada Inc., Swur Son and TNO (Thissen). In the area of direct systems, the firms currently involved include Calfran International, Inc., Calyxes Research and Development Corp., and Freeze Technologies Corp. Some of the current manufacturers are not actively developing or promoting freeze technologies to any great extent; they are primarily holding back until the market is read.

COMPETING TECHNOLOGIES

The predominant competing technologies are evaporation and distillation, though there are other competing technologies depending on the application. Other separation technologies include reverse osmosis (RO) ultrafiltration (UF), gas separation, electrodialysis and ion exchange. The benefits and costs of freeze concentration compared to evaporation or distillation have been discussed in previous sections. Both RO and UF separation systems are also currently being developed as alternatives to evaporation. Both reverse osmosis and ultrafiltration utilize some type of semi-permeable membrane under pressure to separate both organic and inorganic compounds that have been dissolved in aqueous solutions, as well as suspended solids. The main difference between RO and UF separation is the particle size they can filter. RO filters particles that are in the ionic range (below 0.0001 micron), while UF filters particles in the molecular range (0.001 to 0.1 micron).

FUTURE

In general, the long-term market potential for the freeze concentration technology is promising, though for the present and the next few years, freeze concentration growth is likely to be slow due to the stabilization of food industry applications and the development stage of new applications for the pulp and paper, chemical and petroleum industries. There is a great need for pilot and demonstration scale testing in order to convince each industry to adopt freeze technology. This is because the currently used processes are generally well developed and the industries would have to accept the risk of trying a "new" process. The freeze concentration industry generally agrees that freeze concentration will find acceptance only in selected niche markets, such as treating hazardous wastes.

Application for freeze concentration with long-term potential may be in treating industrial and hazardous waste. Because the whole area of environmental issues has become a national concern (to the public, government and industry), there will be significant efforts to find cost effective solutions for hazardous waste treatment, particularly when federal regulations become more strict and enforced.

8 WATER DISINFECTION

THE SCOPE OF THE PROBLEM

Waterborne Diseases

Untreated waters contain a number of harmful pollutants which give the water color, taste and odor. These pollutants include viruses, bacteria, organic materials and soluble inorganic compounds, and these must be removed or rendered harmless before the water can be used again to prevent waterborne diseases.

Typically, breakdown of documented problem outbreaks identifies acute gastroenteritis as responsible for 47% of cases; hepatitis follows with 13%, then shigellosis with 13%, ciardiasis with 12%, chemical poisoning with 8%, typhoid fever with 4% and salmonellosis with 2%. Sources of contaminated water can be traced to semipublic water systems in 55% of cases; municipal water systems implicated in 31% of cases and the remainder (14%) ascribed to individual water systems.

In cell culture, it has been shown that one virion can produce infection. In the human host, because of acquired resistance and a variety of other factors, the one virion-one infection possibility does not exist.

Very little is known of the epidemiology of waterborne diseases. The current data base is insufficient to determine the scope and intensity of the problem. The devastating effect of epidemics is sufficient to rank water-associated epidemics as a most important public health problem.

Viruses and bacteria may be eliminated by chemical methods or by irradiation, and organic poisons may also be controlled. Inorganic matter must be removed by other means.

Characteristics of Viruses

Viruses are ultramicroscopic organisms. They are parasites; they need to infest a host in order to duplicate themselves. Viruses excreted with human and animal feces are called enteric viruses, and more than 100 such organisms have been identified. As many as one million viruses can be found in one gram of excrement. The concentration in raw sewage varies over a wide range; as many as 463,500 infectious particles per liter of raw sewage have been detected.

Origin of Virus

Viruses found in surface waters are introduced from three major sources. Viruses of human origin can be traced to untreated or inadequately treated domestic sewage. Runoffs from agricultural land, feedlots and forests introduce viruses from domestic and wild animals and birds. Plant viruses, insect viruses and other forms of life associated with the aquatic environment may also infect the waters.

Characteristics of Bacteria

In addition to viruses, bacteria (microscopic organisms that can reproduce without a host, given the proper conditions) are also found in water. In general, damage to the human body from bacterial infection is due to the action of the toxins they produce. Bacteria found in water are derived from contact with air, soil, living and decaying plants and animals, and animal excrements.

Many of these bacteria are aerobic and anaerobic spore-forming organisms associated with varying densities of coliforms, fecal coliforms, fecal streptococci, staphylococci, chromogenic forms, fluorescent strains nitrifying groups, iron and sulfur bacteria, proteus species and pathogenic bacteria. Many bacteria are of little sanitary significance and die rapidly in water.

Fecal pollution adds a variety of intestinal pathogens. The most common general found in water are salmonella, shigella, vibrio, mycobacterium, pasturella and leptospira.

Viruses in Water

The circumstances under which water becomes contaminated are as varied as the ways water is taken internally. It is then conceivable that almost any virus could be transmitted through the water route.

The increased use of water for recreational purposes increases the incidence of human contact with bodies of water and, consequently, with waterborne viruses and bacteria.

The major waterborne viruses among pathogens, and the most likely candidates for water transmission, are the picornaviruses (from pico, meaning very small, and RNA, referring to the presence of nucleic acid). The characteristics of picornaviruses are listed in Table 1.

Among the picoviruses are the enteroviruses (polioviruses, coxsackie-viruses and echoviruses) and the rhinoviruses of human origin. Also included are enteroviruses from excrements of cattle, swine and other domesticated animals; and rhinoviruses of non-human origin, viruses of foot-and-mouth disease, teschen disease, encephalomyocarditis, mouse encephalomyelitis, avian excephalomyelitis and vesicular exantherm of pigs.

Additionally, certain viruses can be transported by the water route because their vectors, water molds and nematodes, live in the soil and move with the movement of water. Plant pathogenic viruses also enter the water route and contribute to the problem though this area has been

TABLE 1

PICORNAVIRUS CHARACTERISTICS (VERY SMALL RNA VIRUSES)

1. Small spheres about 20-30 μm in diameter.
2. RNA core, icosahedral form of cubic symmetry.
3. Resistant to ether, chloroform and bile salts, indicating lack of essential lipids.
4. Heat stabilized in presence of divalent cations (Molar $MgCl_2$)
5. Enteroviruses separated from rhinoviruses by acid lability of the latter viruses (inactivated at pH 3.0-5.0)

given little attention in the past. Viruses associated with industrial abattoirs, meat packing, food processing, pharmaceutical and chemical operations are also a potential problem.

All enteric viruses occur in sewage in considerable numbers, and recent detection techniques make it possible to find these viruses in almost all streams that receive sewage effluents. Enteric viruses have been isolated from surface waters around the world. Samples collected from tidal rivers in the U.S. contained viruses in 27 to 52% of the cases. The contamination of surface water by enteric viruses appears to be ubiquitous.

Survival of Viruses

A variety of factors are responsible for the survival of viruses in water bodies. The survival of enteric viruses under laboratory conditions and in estuaries varies from a few hours to up to 200 days. Survival in winter is superior to that of summer temperatures.

It is not known exactly what happens to these multitudes of viruses introduced in water bodies. The inability of rhinoviruses to withstand pH changes, temperature fluctuations and the lack of protective covering offered by feces and other organic materials probably makes the water route of minor importance in their transmission. These factors do not affect the enteroviruses, which are stable and persist in water for long periods. Coxsackieviruses, it has been found, are relatively resistant to concentrations of chlorine normally used for disinfection of bacteria in water. Studies have shown that enteric viruses easily survive present sewage treatment methods and may survive in waters for a considerable time.

Sea or estuary water	2-130 days
River water	2->188 days
Tap water	5-168 days
Soil	25-175 days
Oysters	6-90 days
Landfill leachates	7->90 days
Sediments	?

It has been shown that oysters incorporate poliovirus into their tissues even when grown in sea water with only small amounts of virus.

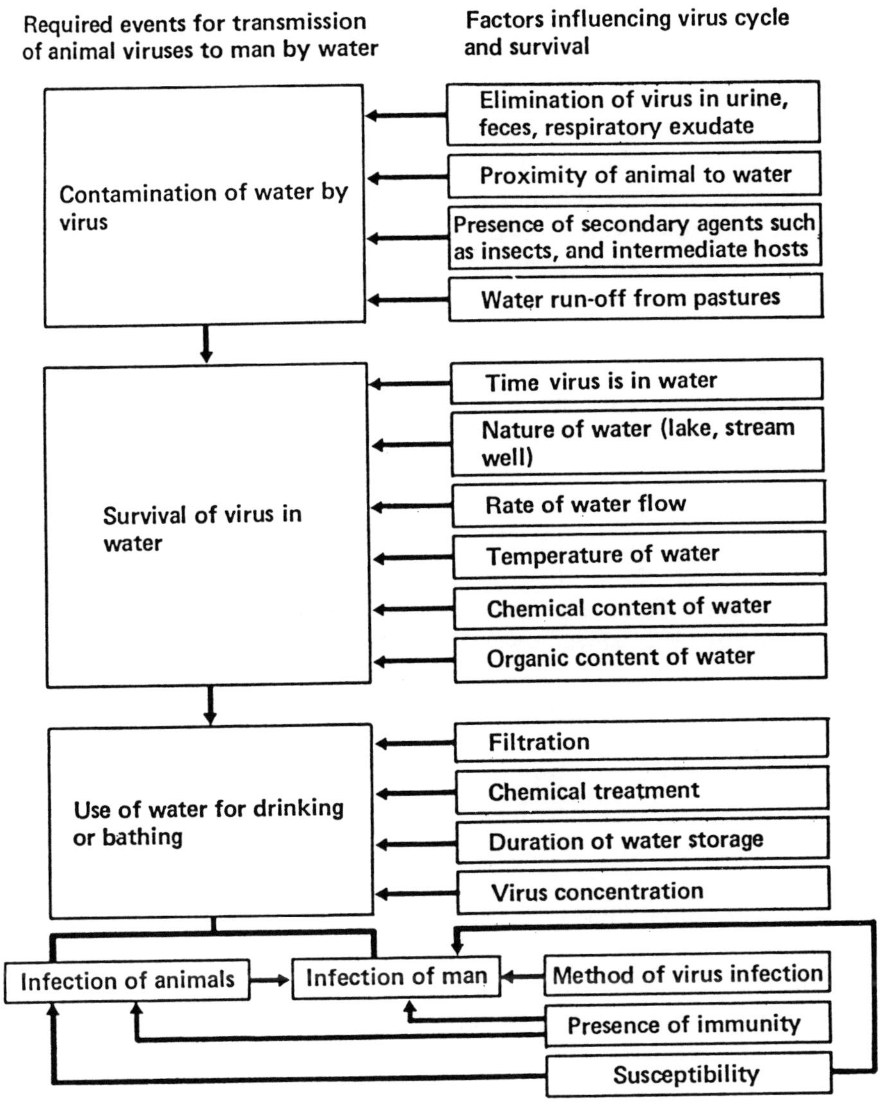

Figure 1. Factors that may affect virus survival in water.

Traditional Treatment Methods

The traditional processes and techniques currently in use for the removal of viruses from water and wastewater include methods effecting physical removal of the particles and those causing the inactivation or destruction of the organism. Among the first included are sedimentation, adsorption, coagulation and precipitation and filtration. The second encompasses pH and chemical oxidation by disinfection such as halogens.

Primary treatment of municipal waste involving settling and retention removes very few viruses. Sedimentation effects some removal. Virus removal of up to 90% (which is a minimal removal efficiency) has been observed after the activated sludge step. Further physical-chemical treatment can result in large reduction of virus titer, coagulation being one of the most effective treatments achieving as much as 99.99% removal of virus suspended in water. If high pH (above 11) is maintained for long periods of time, 99.9% of the viruses can be removed.

Of all the halogens, chlorine at high doses (40 mg/l for 10 min) is very effective, achieving 99.9% reduction. Lower doses (e.g., 8 mg/l) result in no decrease in virus.

As a result of several studies, the following general conclusions regarding viruses in sewage warrant consideration:

- Primary sewage treatment has little effect on enteric viruses.
- Secondary treatment with trickling filters removes only about 40% of the enteroviruses.
- Secondary treatment by activated sludge treatment effectively removes 90 to 98% of the viruses.
- Chlorination of treated sewage effluents may reduce, but may not eliminate, the number of viruses present.

The current concept of disinfection is that the treatment must destroy or inactivate viruses as well as bacillary pathogens. Under this concept the use of coliform counts as an indicator of the effectiveness of disinfection is open to severe criticism given that coliform organisms are easier to destroy than viruses by several orders of magnitude.

An important concept that surfaced in recent years is that a single disinfectant may not be capable of purifying water to the desired degree. Also, it might not be practicable or cost-effective. This realization has given rise to a variety of treatment combinations in series or in parallel. The analysis further indicates that the search for the perfect disinfectant for all situations is a sterile exercise.

It has been estimated that in the U.S. only 60% of municipal waste effluent is disinfected prior to discharge and, in a number of cases, only on a seasonal basis. Coupling this fact with the demonstration that various sewage treatment processes achieve only partial removal of viruses leaves us with a substantial problem to resolve.

FUNDAMENTAL METHODS

Electromagnetic Waves

Electromagnetic radiation is the propagation of energy through space by means of electric and magnetic fields that vary in time. Electromagnetic radiation may be specified in terms of frequency, vacuum wavelength or photon energy. The electromagnetic spectrum with its application to water treatment is shown in Table 2.

For water purification, EM waves up to the low end of the UV band will result in heating the water. (This includes infrared as well as most lasers). In the visible range, some photochemical reactions such as dissociation and increased ionization may take place. At the higher frequencies it will be necessary to have thin layers of water because the radiation will be absorbed in a relatively short distance. It should be noted that the conductivity and dielectric constant of materials are, in general, frequency-dependant. In case of the dielectric constant, it decreases as 1/wavelength. Hence, the electromagnetic absorption will vary with the frequency of the applied field. There may be some anomalies in the absorption spectra in the vicinity of frequencies that could excite molecules. At those frequencies, the absorption could be unusually large.

UV radiation in the region between 0.2μ and 0.3μ has germicidal properties. The peak germicidal wavelength is around 0.26μ (2600Å). This short UV is attenuated in air and, hence, the source must be very near the medium to be treated. Again, the medium must be very thin as the UV will be attenuated in the medium as well.

X-rays and gamma rays are high-energy photons and will tend to ionize most anything with which they collide. They could generate UV

TABLE 2

THE ELECTROMAGNETIC SPECTRUM

Frequency Range	Vacuum Wavelength	Application	Method of Generation
Up to 20 KHz	Down to 10 km	Power & telephone	Rotating generators, microphones, transistors & tubes
20 KHz - 500 MHz	10 km - 0.5 m	Radio & TV	Tubes & transistors
555 KHz - 1605 KHz		AM radio	
54 MHz - 88 MHz		TV Channels 2-6	
88 MHz - 108 MHz		FM radio	
174 MHz - 46 MHz		TV Channels 7-13	
500 MHz - 5000 GHz	15 m - 0.1 mm	Industrial radar	Semiconductors, klystrons
470-890 MHz		TV Channels 14-82	Magnetrons, lasers
5000 GHz - 5 x 10^5 GHz	0.1 mm to 0.7μ	Heat	Lamps, lasers
5 x 10^5 GHz - 3 x 10^{17} Hz	0.7μ - 0.4μ	Light	Lamps, lasers
7.5 x 10^5 GHz - 3 x 10^{17} Hz	0.4μ - 10Å	UV	UV lamps, spark discharge, lasers, laser-doubling
3 x 10^{17} Hz - 3 x 10^{20} Hz	10Å - 0.1Å	X-rays	X-ray tubes, radioactive isotopes
3 x 10^{20} Hz - 3 x 10^{24} Hz	0.01Å - 0.001mÅ	Gamma rays	Linear accelerators, nuclear reactions
3 x 10^{24} Hz - Up	0.001 mÅ & shorter	Cosmic rays	Cosmic sources

in air. At higher energies it is possible for the gamma rays to induce nuclear reactions by stripping protons or neutrons from nuclei. This could result in the production of isotopes and/or the production of new atoms.

Sounds

A sound wave is an alteration in pressure, stress, particle displacement, particle velocity or a combination of the above that is propagated in an elastic medium. Sound waves, therefore, require a medium for transmission, i.e., they may not be transmitted in vacuum.

The sound spectrum covers all possible frequencies. The average human ear responds to frequencies between 16 Hz and 16 KHz. Frequencies above 20 KHz are called ultrasonic frequencies. Sound waves in the 50-200 KHz range are used for cleaning and degreasing.

In water purification applications, ultrasonic waves have been used to effect disintegration by cavitation and mixing of organic materials. The waves themselves have no germicidal effect, but, when used with other treatment methods, can provide the necessary mixing and agitation for effective purification.

Electron Beams

The electron is the lightest stable elementary particle of matter known and carries a unit of negative charge. It is a constituent of all matter and can be found free in space.

Electron Constants

Mass	0.109×10^{-31} kg
Charge	1.602×10^{-19} Coulomb
Spin	0.5

Under normal conditions, each chemical element has a nucleus consisting of a number of neutrons and protons, the latter equal in number to the atomic number of the element. Electrons are located in various orbits around the nucleus. The number of electrons is equal to the number of protons, and the atom is electrically neutral when viewed from a distance. The number of electrons that can occupy each orbit is governed by quantum mechanical selection rules. The binding energy between an electron and its nucleus varies with the orbit number, and in general, the electrons with the shortest orbit are the most tightly bound. An electron can be made by giving it a quantum of energy. This energy quantum is fixed for any given transition and whether a transition will occur is again governed by selection rules. In other words, although an electron is given a quantum of energy sufficient to raise it to an adjacent higher state, it will not go up to that state if the transition is not permitted. In that case, it is theorized that if the electron absorbs the quantum, it will most probably go up to the excited state, remain there for a time allowed by the Uncertainty Principle, reradiate the quantum and return to its original state.

If an electron is given a sufficiently large quantum of energy, it will completely leave the atom. The electron will carry off as kinetic energy the difference between the input quantum and the energy required to ionize. The remaining atom will now become a positively charged ion, and the stripped electron will become a free electron.

This electron may have sufficient energy when it leaves the atoms (or it may acquire sufficient energy from an external field) to collide with another atom and strip it of an electron. This is the basis for electric discharge where free electrons are accelerated by an applied field and, as they collide with neutral atoms generate additional free electrons. This process avalanches as the electron approach the positive electrode. At the same time, the positively charged ions are accelerated towards the negative electrode.

In vacuum, when a voltage is applied between two electrodes, electrons will move from the cathode to the anode. Of course, in vacuum there will be no avalanching effects. Electrons are emitted from the cathode by a number of mechanisms:

- **Thermionic Emission**--Because of the non-zero temperature of the cathode, free electrons are continuously bouncing inside. Some of these have sufficient energy to overcome the work function of the material and can be found in the vicinity of the surface. The cathode may be heated to increase this emission. Also to enhance this effect, cathodes are usually made of, or coated with, a low work-function material such as thorium.

- **Shottky Emission**--This is also a thermionic type of emission except that in this case, the applied field effectively decreases the work function of the material, and more electrons can then escape.
- **High Field Emission**--In this case, the electric field is high enough to narrow the work-function barrier and allow electrons to escape by tunnelling through the barrier.
- **Photoemission**--EM radiation of energy hf can cause photoemission of electrons whose maximum energy is equal to or larger than the difference between the photon energy and the work function of the material.
- **Secondary Emission**--Electrons striking the surface of a cathode could cause the release of some electrons and, hence, a net amplification in the number of electrons. This principle is used in the construction of photo-multipliers where light photons strike a photoemitting cathode releasing photoelectrons. These electrons are subsequently amplified striking a number of electrodes (called dynodes) before they are finally collected by the anode.

Electromagnetism

In a high-gradient magnetic separator, the force on a magnetized particle depends on the intensity of the magnetizing field and on the gradient of the field.

When a particle is magnetized by an applied magnetic field, the particle develops an equal number of north and south poles. Hence, in a uniform field, a dipolar particle experiences a torque, but not a net tractive force. In order to develop a net tractive force, a field gradient is required; that is, the induced poles at the opposite ends of the particle must view different magnetic fields.

In a simplified, one-dimensional case, the magnetomotive force on a particle is given by

$$F_m = \mu(\delta H/\delta x) = MV\ (\delta H/\delta x) = xHV\ (\delta H/\delta x)$$

where μ is the magnetic moment of the particle under field intensity, H. $\delta H/\delta x$ is the field gradient. The magnetic moment μ is the product of the magnetization of the particle and its volume (μ = MV). And magnetization is the product of the particle susceptibility, x, and the field intensity, H. In water purification, this magnetic force may be used to separate magnetizable particles.

Direct and Alternating Currents

Electrolytic treatment is achieved when two different metal strips are dipped in water and a direct current is applied from a rectifier. The higher the voltage, the greater the force pushing electrons across the gap between the electrodes. If the water is pure, very few electrons cross the path between the electrodes. Impurities increase conductivity, hence decreasing the required voltage. Additionally, chemical reactions occur at both the cathode and the anode. The major reaction taking place at the cathode is the decomposition of water with the evolution of hydrogen gas. The anode reactions are oxidations by four major means:

1. Oxidation of chloride to chlorine and hypochlorite.
2. Formation of highly oxidative species such as ozone and peroxides.
3. Direct oxidation by the anode.
4. Electrolysis of water to produce oxygen gas.

APPLICATIONS

Electrolytic Treatment

A great deal of interest was generated in the U.S. prior to 1930 in electrolytic treatment of wastewater, but all plans were abandoned because of high cost and doubtful efficiency. These systems were based on the production of hypochlorite from existing or added chloride in the wastewater system. A great deal of interest has surfaced in re-evaluating such techniques.

In a series of recent experiments using both direct and alternating current, the emphasis was placed on disinfection, especially the inactivation or killing of three bacteria: Escherichia coli, Pseudomonas aeruginosa and Klebsiella pneumoniae; two viruses: polio 1 and coxsackie B; one yeast: Saccharomyces cerevisiae; and the normal flora of microorganisms found in domestic sewage.

The critical parameters studies include:

1. The peak value of applied voltage (4 to 19V).
2. The resulting current density (10 to 55 mA/cm^2).
3. The wave form (sine, triangular, rectangular).
4. The frequency of the applied voltage (0.5 Hz to 1.5 Hz).
5. The process water flow rate (100 to 500 cm^3/min).
6. The alkalinity/acidity of the water (pH 5 to 9).
7. Turbidity (0 to 50 JTU).
8. Presence or absence of sodium chloride.
9. Water conductivity (120 to 3000 μmho/cm).
10. Reactor exposure time (0 to 6 min).
11. Flow regime (laminar or turbulent).

In general, it was found that alternating current gave better results than direct current. It has also been suggested that the superimposition of a high frequency a.c. on the d.c. would reduce the work function and would cause a promotion of electron transfer at the electrode interface.

The electrochemical biocide is compared in an indirect fashion with traditional disinfection techniques. The results indicate that chemical techniques are currently more effective in destroying organisms at the current stage of development. Increased efficiency must be achieved for the electrolytic technique to compete in large-scale systems with chemical agents.

While the process is not fully elucidated, it appears that four mechanisms are involved:

- **Reduction Number of Viable Microorganisms by Adsorption onto the Electrodes**--Protein and microorganisms adsorption on electrodes with anodic potential has been documented. Microorganism adsorption on passive electrodes (in the absence of current) has been observed with subsequent electrochemical oxidation. This does not appear to be a major route for inactivation.
- **Electrochemical Oxidation of the Microorganism Components at the Anode**--Oxidation of various viruses due to oxidation at the surface of the working electrode has been indicated, although the peak voltage used in many

experiments would not be sufficient for the generation of molecular or gaseous oxygen.

- **Destruction of the Microorganisms by Production of a Biocidal Chemical Species**--It has been shown that NaCl is not needed for effective operation in the destruction of microorganisms. Biocidal species such as Cl^-, HO^-, $O^=$, ClO^- and HOCl occur but have very low diffusion coefficients. Hence, if this phenomenon occurs, the probability is that organisms are destroyed at the electrode surface rather than in the bulk solution.
- **Destruction by Electric Field Effects**--It has been observed that some organisms are killed in midstream without contact with the electrodes. The organisms were observed to oscillate in phase with the electric field. Hence, microorganism kill can also be ascribed to changes caused by changing electromotive forces resulting from the impressed a.c.

Electromagnetic Separation

In the typical operation, a magnetized fine-particle seed (typically iron oxide) and a flocculent (typically aluminum sulfate) are added to the wastewater, prompting the formation of magnetic microflocs.

The stream then flows through a canister packed with stainless steel wire and a magnetic field is applied. The stainless steel wool captures the flocs by magnetic forces.

Ozonation

Ozone has been used continuously for nearly 80 years in municipal water treatment and the disinfection of water supplies. It started in France, then extended to Germany, Holland, Switzerland and other European countries, and in recent years to Canada.

Ozone is a strong oxidizing substance with bacterial properties similar to those of chlorine. In test conditions it was shown that the destruction of bacteria was between 600 to 3000 times more rapid by ozone than by chlorine. Further, the bacterial action of ozone is relatively unaffected by changes in pH while chlorine efficacy is strongly dependent on the pH of the water.

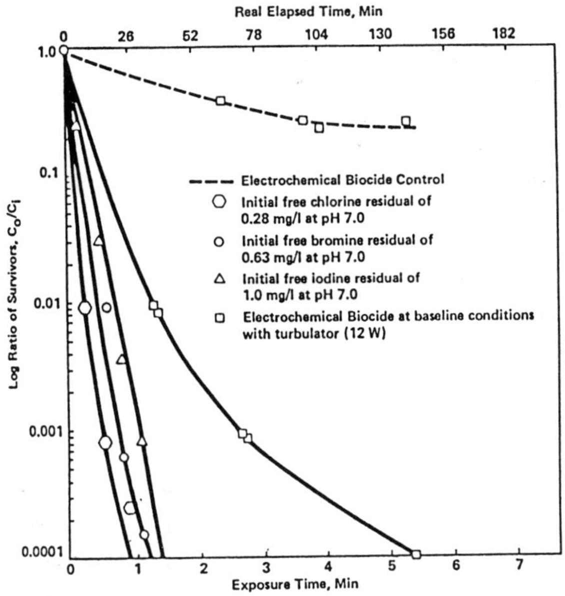

Figure 2. Microorganism survival ratios versus different disinfecting agents.

Ozone's high reactivity and instability as well a serious obstacles in producing concentrations in excess of 6% preclude central production and distribution with its associated economies of scale.

In the electric discharge (or corona) method of generating ozone, an alternating current is imposed across a discharge gap with voltages between 5 and 25 kV, and a portion of the oxygen is converted to ozone. A pair of large-area electrodes are separated by a dielectric (1-3 mm in thickness) and air gap (approximately 3 mm). Although standard frequencies of 50 or 60 cycles are adequate, frequencies as high as 1000 cycles are also employed.

The mechanism for ozone generation is the excitation and acceleration of stray electrons with the high voltage field. The alternating current causes the electron to be attracted first to one electrode and then to the other. As the electrons attain sufficient velocity, they become capable of splitting some oxygen molecules into free radical oxygen atoms. Those atoms may then combine with 0_2 molecules to form 0_3.

The principal ozone decomposition products in aqueous solution are molecular oxygen and the highly reactive free radicals HO^2, OH^-, and H^+.

At present, more than 1000 treatment plants around the world ozonate drinking water. The cost of using ozone for disinfection is expected to be around $0.07/1000 gal and involves the consumption of 10 kWh of electricity per pound of ozone generated.

Canada has 11 water treatment plants using ozone and has built the largest plant in the world, 600 mgd (a large facility is in Moscow with 300-mgd capacity).

A study on the combined effect of ozone and sound on the inactivation of several pathogenic and non-pathogenic bacteria in secondary wastewater effluent has been reported. The treatment enhanced the rate of inactivation of all bacteria strains.

Besides the disinfection of sewage effluent, ozone is used for sterilizing industrial containers such as plastic bottles, where heat treatment is inappropriate. Breweries use ozone as an antiseptic in destroying pathogenic ferments without affecting the yeast. It is also used in swimming pools and aquariums. It is sometimes used in the purification and washing of shellfish and in controlling slimes in cooling towers. Ozone has also been shown to be quite effective in destroying a variety of refractory organic compounds.

Ultraviolet Radiation

It has been shown that:

- Ultraviolet radiation around 254 mm renders bacteria incapable of reproduction by photochemically altering the DNA of the cells.
- 99% of fecal coliform and fecal streptococcus can be killed by a fairly low dose of ultraviolet light.
- Bacterial kill is independent of the intensity of the light but depends on the total dose.
- Simultaneous treatment of water with UV and ozone results in higher microorganism kill than independent treatment with both UV and ozone.
- When ultrasonic treatment was applied before treating with the UV light, a higher bacteria kill was obtained.

- The UV dose required to reduce the survival fraction of total coliform and fecal streptococcus to 10^{-2} (99% removal) is approximately 4 x 10^{-10} Einsteins/ml.
- Rough laboratory cost estimates indicate a cost of \$0.002 to \$0.004/m^3 for a UV dose of 4 x 10^{-8} Einsteins/ml.

Some limitations are associated with UV radiation for disinfection:

- The process performance is highly dependent on the efficacy of upstream devices that remove suspended solids.
- Another key factor is that the UV lamps must be kept clean in order to maintain their peak radiation output.
- A further drawback is associated with the fact that a thin layer of water (<0.5 cm) must pass within 5 cm of the lamps.

One way of implementing the UV disinfection process at existing activated sludge plants involves suspending the UV lights (in the form of low-pressure mercury arc UV lamps with associated reflectors) above the secondary clarifiers. The effluent is exposed to the UV radiation as it rises over the wire in a thin film.

Using the above arrangement, the cost (capital and operating) for a pilot 1-mgd plant would be, respectively, \$12,500 and \$0.018/1000 gal (Imp.). A full-scale facility would cost between \$0.02 and \$0.05/1000 gal to operate.

A comparison has been made of the comparative effectiveness of chlorine, ozone and ultraviolet light in inactivating viruses and bacteria. Some of the bacteria and viruses are much less sensitive to chlorine and ozone than fecal coliform. On the other hand, in all cases, the UV dosage requirements relative to coliform are one or less for all organisms studied, indicating that all pathogens and viruses studies are as sensitive or more sensitive to UV light than are fecal coliforms. Thus, if sufficient UV radiation is applied to kill 99% of all fecal coliforms, one can expect a 99% kill of these pathogens and viruses as well.

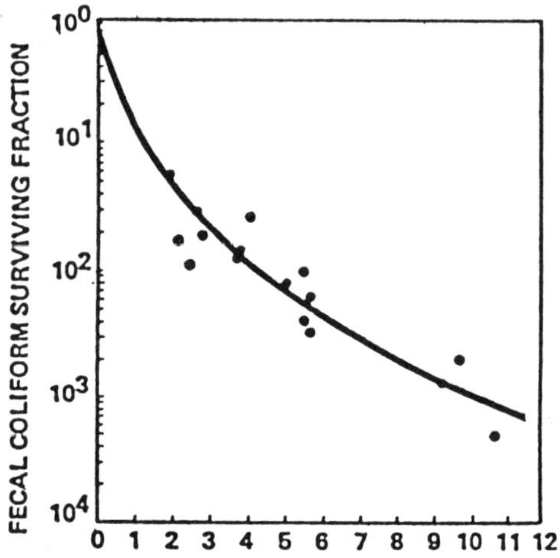

Figure 3. Ultraviolet dose (Einsteins/ml x 10^{-8}).

Electron Beam

The idea of using ionizing radiation to disinfect water is not new. Ionizing radiations can be produced by various radioactive sources (radioisotopes), by x-ray and particle emissions from accelerators and by high-energy electrons. The advances in reliable, relatively low-cost devices for producing high-energy electrons are more significant.

Unlike x-rays and gamma rays, electrons are rapidly attenuated. The maximum range of a 1-million-volt electron is about 4 m in air and about 5 cm in water.

In transit in matter, an electron loses energy through collisions that ionize atoms and molecules along its path. Bacteria and viruses are destroyed by the secondary ionization products produced by the primary traversing electron.

The energetic electrons dissociate water into free radicals H^+ and OH^-. These may combine to form active molecules-hydrogen peroxides and ozone. These highly active fragments and molecules attach to living structures to promote their oxidation, reduction, dissociation and degradation.

Previous studies have indicated that 400,000 rads would be adequate for sewage disinfection. At 100 ergs per gram rad, 400,000 rads would raise the temperature of the water or sludge by 1°C. At this dose, each cm^2 of moving sludge would receive about 12 x 10^{12} electrons, each electron producing some 300,000 secondary ionizations.

Shown is a survival test organisms as a function of dose. The presence of oxygen further enhances the lethal effect of the electrons.

This concept has been applied at the Deer Island wastewater treatment plant near Boston, Massachusetts, by the Massachusetts Institute of Technology (M.I.T.) and High Voltage Engineering of Boston, Massachusetts.

Using an off-the-shelf 50-kW, 750-kV electron beam, it is possible to deliver 400 Krad to the thickened sludge. The electrons are accelerated in vacuum. They come out through a titanium window and impinge upon the thin layer of sludge. The titanium window is about 15 cm above the surface of the rotating drum. The sludge is rolled onto a steel drum in a thin layer 2 mm thick by 1.2 m wide and travels at 2 m/sec through the electron beam. The electrons travel in air approximately 15 cm where they generate ozone which contributes to the treatment's procedure. The capacity is 70 gpm or 100,000 gpd. The time of transit is such that any increment of sludge will pass through the intense electron region in one hundredth of a second.

The total dose of about 400,000 rads kills all the bacteria and living organisms. It has also been found that the radiation dissociates the complicated molecules of the pesticides such as TCB and kepone and converts them into their natural constituents. In addition, the radiation reduces the smell of the sludge. For bacteria and virus disinfection, the process is considered successful at present. The Deer Island facility went into intermittent operation in April 1976 and is designed to treat 100,000 gallons of sludge per day.

The estimated capital cost for a 100,000-gpd system was to be $500,000 and the operating cost, $100,000 per year or $3.00/100 gal.

Gamma Radiation

No operating gamma ray water disinfection system exists in the U.S. Experimental tests have been conducted.

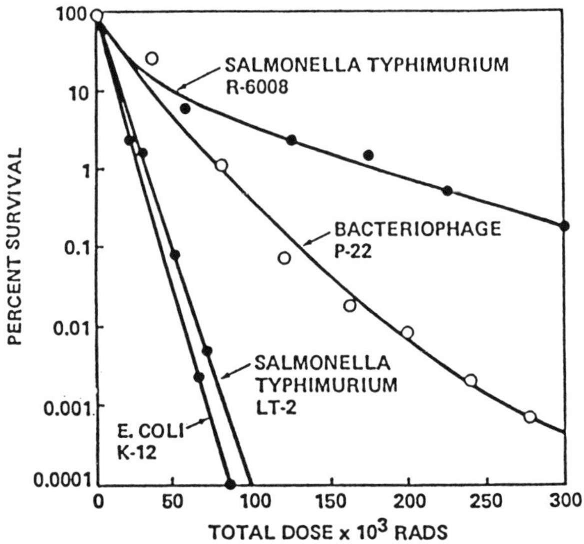

Figure 4. Survival of test organism as a function of dose.

	Chlorination	Irradiation
Coliforms	98.2%	97.9%
Fecal Coliforms	99.8%	83.9%
Fecal Streptococcia	98.8%	46.0%

Detailed analyses have shown that 300 krad of absorbed radiation is equivalent to that of a heat treatment at 80°C for 30 min. This kind of treatment results in a reduction by a factor of 10^6 to 10^7 of the population of intestinal microorganisms.

Using a cobalt-60 unit containing 120 kci, a sludge sterilization program was initiated in Germany. Using an irradiation time of 210 min giving average absorbed dose of 318 krad, a sludge containing 1.4 to 4.5% solid was successfully sterilized. The total operating costs are on the order of 2.0 dm/m^3 of sewage sludge, and it would be around 1.2 dm/m^3 for a Cs source. The electric power consumption for the total systems was on the order of 5.2 kWh/m^3 sludge.

By comparison, pasteurization with steam costs around 1.84 dm/m^3 with an increase in volume of 20%. The gamma ray sterilization system actually dewaters the sludge partially. In 30 days, 30% water was lost.

A combination of gamma radiation, ozonation and carbon adsorption has been applied to treating municipal waste streams with organic concentrations in the range of 100 to 1150 ppm with oxidation of 95% of the organic load. The cost was about 28¢/1000 gal for a 500,000 gal/day unit. The Florida Institute of Technology has reported that gamma ray treatment coupled with heating the water or bubbling air through it during irradiation is effective for disinfection.

Microwave

Microwaves have been used for dewatering in a number of applications (food concentration, high-quality paper drying). It is not surprising that attempts are made to use microwaves for sludge dewatering.

Microwaves have been used in one application to sterilize solids and ready them for a disposal. The Watertek system, combining microwaves and ozone, is a factory-assembled system available with capacities from 30,000 to 50,000 gal/day. Treating soils with microwaves eliminates the need for settling ponds, digester and sludge beds and saves considerable space and time. With a 60-min residence, the designer claims a solid output, free of disease-causing agents and odors and capable of meeting the Environmental Protection Agency's requirements for wastewater effluents. Operating costs are $0.256/1000 gal of raw sewage, based on electric power at $0.03/kWh.

Laser

We have not found any application of laser technology for disinfection, but it could conceivably be used for that purpose. When a laser beam impacts on biological material, a variety of physical and chemical reactions may take place depending on the wavelengths of the laser and the power density used. For wavelengths in the red and infrared region, the main effect is thermal, the absorbed energy being converted into heat.

Laser illumination in the ultraviolet range may also photochemically excite molecules besides the thermal effect. Lasers can initiate photodynamic action in biological materials. This process is enhanced by the irradiation of dye sensitizers. These compounds may form long-life metastable complexes where the energy from the absorbed photon is available to catalyze biochemical reactions.

TABLE 3

WATER DISINFECTION ECONOMICS

	Plant size, Million Gallons per Day					
	1	**10**	**100**	**1**	**10**	**100**
	Capital Cost $1000			**Operating Costs ¢/1000 gallons**		
Chlorine	60	190	840	3.5	1.4	0.7
Chlorine/activated	640	2800	8400	19.0	8.6	3.3
Ozone (from air)	190	1070	6880	7.3	4.0	2.8
Ozone (from oxygen)	160	700	4280	7.2	3.5	2.4
Ultraviolet Radiation	70	360	178	4.2	2.7	2.3
Bromine Chloride	50	130	410	4.5	3.0	2.7
Electrolytic				310.0		
Electron Beam	500			300.0		
Gamma Radiation						
Electromagnetic		500-1000			5-10	
Microwave				26.0		

A practical view of the future allows the following conclusions:

- Interest will continue to grow in achieving higher levels of disinfection.
- It is useless to look for the perfect disinfection system.
- Chlorine, because of the production of chlorinated organics will be severely re-evaluated as the preferred method of disinfection.
- Ozone is a likely candidate for replacing chlorine and is competitive with the chlorination-dechlorination reaeration system.
- Electrotechnologies, in general, need to achieve higher efficiency and reduced costs.
- Electrotechnologies have not been applied on a large scale as yet.

INDEX